COMPETITIVE MANUFACTURING

Hal Mather

PRENTICE HALL
Englewood Cliffs, New Jersey 07632

Prentice-Hall International (UK) Limited, *London*
Prentice-Hall of Australia Pty. Limited, *Sydney*
Prentice-Hall Canada, Inc., *Toronto*
Prentice-Hall Hispanoamericana, S.A., *Mexico*
Prentice-Hall of India Private Limited, *New Delhi*
Prentice-Hall of Japan, Inc., *Tokyo*
Simon & Schuster Asia Pte. Ltd., *Singapore*
Editora Prentice-Hall do Brasil, Ltda., *Rio de Janeiro*

© 1988 by

PRENTICE-HALL, Inc.

Englewood Cliffs, NJ

Printed in the United States of America

10 9 8 7 6 5 4 3 2 1

Library of Congress Cataloging-in-Publication Data

Mather, Hal.
    Competitive manufacturing / Hal Mather.
        p.   cm.
    Includes index.
    ISBN 0-13-155029-2
    1. Production management.   I. Title.
TS155.M3393   1988
658.5—dc19   88-716
                                                         CIP

ISBN 0-13-155029-2

ISBN 0-13-156753-5 PBK

10 9 8 7 6 5 4 3 2   PBK

PRENTICE HALL
BUSINESS & PROFESSIONAL DIVISION
A division of Simon & Schuster
Englewood Cliffs, New Jersey 07632

About the Author

Hal Mather is president of HAL MATHER, INC., Atlanta, GA, an international management consulting and education company. Since 1973 he has been helping all types of industrial concerns to improve their business planning and control. Recent assignments have taken him throughout North America, Europe, the Far East, Australasia, South Africa, Mexico, and Brazil.

Mr. Mather helps companies two ways, the first of which is personal consulting. He has worked with both large and small companies and has stimulated many successful projects with enormous payback.

The second way he helps companies is as a dynamic lecturer and educator. He conducts private courses for companies, tailored specifically to their products, processes, and problems. His expertise encompasses MRP II, JIT, and CIM, and he is a leading expert in the factory of the future. His educational style motivates the key managers in a business to make the quantum changes necessary to make a company a world-class competitor.

Mr. Mather is a prolific author. His many articles have appeared in a number of magazines, among them the *Harvard Business Review* and *Chief Executive*, and he has been quoted in *Fortune, Inc.*, and *Industry Week*. He won the Romeyn Everdell award in 1987 for the best article published in the *Production and Inventory Management Journal*. His two books, *Bills of Materials, Recipes and Formulations* and *How to Really Manage Inventories*, are classics in the field.

Hal Mather has been certified at the Fellow level by the American Production and Inventory Control Society, is a Fellow of the Institution of Mechanical Engineers (U.K.), a Senior Member of the Computer and Automated Systems Association of the Society of Manufacturing Engineers, and is a member of both the Institute of Industrial Engineers and the Association for Manufacturing Excellence. He is listed in Who's Who in the South and Who's Who in Finance and Industry.

Other Books
by the Author

Bills of Materials, Dow Jones-Irwin, 1987.

How to Really *Manage Inventories*, McGraw-Hill, 1984.

What This Book Will Do for You

Competitive Manufacturing is a major breakthrough in the dynamic, rapidly changing field of operations planning and control. Internationally known manufacturing pioneer Hal Mather shows you how to run your business the *right* way to fatten your bottom line, sharpen your competitive edge, and take the lead in the race toward the factory of the future.

While operations is its starting point, *Competitive Manufacturing* goes a giant step beyond the shop floor. It has radical implications for every function in your company—sales, finance, and production included—and every employee from the lunchroom to the boardroom. Here are just a few of the reasons why *Competitive Manufacturing* should be required reading for every senior and middle manager:

- It shows you how to produce the right products for the right markets at the right time, so you can cut inventories to the bone, free up space for new ventures, and generate additional cash.
- It explains how to streamline and simplify factory operations to make your labor more productive and your business easier to manage.
- It tells you how to build the products customers want and deliver them right on time, giving you a leg up on your competitors—both domestic and foreign.
- It shows you how to put some teeth into your long-range planning. Mather demonstrates how strategic planning can be more than just an idle exercise; it becomes a vital force that drives the total business.
- It spells out the right way to design new products to eliminate two problems that plague manufacturers everywhere: critical shortages of some items and excess inventories of others. You'll also see how to manage a varied product line and overcome the single biggest obstacle to effective management.
- It reveals why the traditional accounting techniques you've been

using are dead wrong and dangerous to the health of your business. You'll see how to use operation-based measures to take your company's pulse—and make smarter management decisions.

- It introduces the latest developments in manufacturing management and technology, including just-in-time, computer integrated manufacturing, and the automated factory. You'll see how other companies are putting these concepts to work—and what you can do today to pave the way for the factory of tomorrow.

In simple, practical terms, *Competitive Manufacturing* shows you how to come to grips with the most nagging problems of industrial management and how to get every department to work in concert to benefit the total business. Your bottom-line results will be nothing less than total enterprise excellence.

Contents

Preface

Executives in manufacturing and in industry in general are woefully uninformed about a key aspect of their business: how to achieve excellence in operations. They are knowledgeable about most other aspects—for example, sales, research and development, finance and the actual manufacturing process—but the management of operations is the weakest link in most companies.

Few executives have spent enough time in the operations planning and control area during their careers to develop a good understanding of it. What knowledge they do have is based on observation of the way the business is actually run. This is like studying the local joggers when you wish to become an Olympic miler.

Operations planning and control, or what I call "logistics," can be a confusing area. Huge amounts of data are used, the practitioners speak a language all their own, and things rarely work out the way they were planned. For example, a normal six-week lead-time item is produced on the night shift if hot enough. So what's the real lead time, one day or six weeks? Then a seemingly simple job takes months to complete.

I am going to explain the logistics side of industry in simple terms and show how it really should be managed. I'll debunk many myths around the subject and introduce you to some of the latest concepts, such as Just-in-Time, Computer Integrated Manufacturing, and the automated factory of the future.

My main objective, though, is to show you how to manage logistics to increase return on investment (ROI). Logistics directly influences inventory levels, customer service, and factory productivity. Improvements in this area can reduce inventories at the same time as profits are increased. This double attack on the return on investment equation cannot be claimed by any other activity. Doubling your ROI in 12 months should be possible after reading this book and applying its concepts.

This will not be possible if you leave this area to the "experts." Logistics is heavily dependent on, and in turn heavily influences, other business functions. A team effort from sales, accounting, engineering, and manufacturing will be necessary to make real progress.

Make sure the heads of all these functions read pertinent chapters of this book. The knowledge they gain will give them a clear understanding of what their function must do to get the total business results that are possible.

As we move to the automated factory of the future, logistics will increase in importance. The technical problems of an automated factory (flexible machining systems, robotics, and computer aided design), are already well defined and will be solved.

The tough problem ahead is the management integration of all this technology. What tasks will we set for the business? Are new policies and procedures needed in this changed world? How will we optimize the use of these costly investments? What degree of flexibility do we need? These are just some of the management questions that must be addressed.

It is also obvious that the simpler and more predictable operations are, the sooner the factory can be automated. The race is on to be the first with an automated factory and this race is an international one, not a domestic one.

Computer Integrated Manufacturing is just around the corner, offering untold opportunities to be more competitive. But as with all new technology, there will be winners and losers. The company that is first to take advantage of this new technology and develop a factory of the future will end up having a factory *with* a future.

TERMINOLOGY

I find it difficult to write clearly about manufacturing companies because the word "manufacturing" often has several meanings. It can mean that business discipline concerned with making the product, it can mean an individual manufacturing company, it can mean a whole industrial group, or it could even define that segment of our economy concerned with making things. There's obviously a vast difference in these meanings.

I'll try to follow a standard format to eliminate confusion. "Manufacturing" will be used to describe that business discipline that is concerned with making the product. It will usually include supporting disciplines such as industrial engineering, materials management, and purchasing. I'll use the terms "manufacturer" or "industrial concern" to denote a company in the manufacturing business with all business disciplines, such as sales, accounting, and engineering, included.

When it comes to talking about our economy, things get tougher, because many government statistics are published about "manufacturing." I'll write "manufacturing segment" or "industrial sector" to denote this portion of our gross national product (GNP), as distinct from banking, transportation, or insurance.

ABOUT THE AUTHOR

Since 1973, I have been working with industrial companies around the world in a dual role. Half my time is spent consulting on operations problems. I spend the other half conducting senior and middle management courses on various aspects of the operations arena. Twenty years' experience in a variety of positions in industry prior to 1973 gave me the background for these activities. Consulting, lecturing, and studying the field since that time have broadened my view of the operations area.

It's from this base that this book has been drawn. My ideas have developed from real-life successes and failures of a variety of approaches and from my association with some key pioneers in the field, the most prominent being George Plossl, a well-known consultant, lecturer, and author.

The primary reason for this book is the surprised reaction of senior- and middle-level managers, from manufacturing companies around the world, to well-known concepts. These reactions are usually expressed while I am explaining these concepts in a course. They seem to be revelations to most attendees.

Typical comments are, "You have opened a window to some new ideas and concepts to resolve some constant, nagging problems, like poor customer service, high inventories, and excess costs, that I never thought of before," or, "I now see what crucial role my function plays in getting more effective operations and hence better business results," or, "I see much clearer now how to integrate the separate business functions." But the most telling comment that drove me to write this particular book is, "Now I see what I must do to become a world-class competitor."

Being the best manufacturer of a given product in America is no longer adequate. Worldwide trade, the growth of industries in previously undeveloped parts of the world, such as the Far East and Latin America, and their need to export manufactured goods to earn foreign currency, has put a different slant on the competitive scene. Your object now must be to compete with the best in the world.

Many changes are on the horizon. The often-promised but so far unfulfilled "factory of the future" is getting closer to reality. Those companies able to implement and manage this new and revolutionary technology will be tomorrow's survivors.

Excellence in operations will be mandatory to run an automated factory. Nothing less will be acceptable. Get your act together now and be ready to capitalize on the new technology as it appears.

Time is running out. The pressures on American industry will not reduce—if anything they will increase. It is my hope that you will implement some of the concepts covered in this book quickly, to put you firmly on the road to becoming a world-class competitor through excellence in operations.

Hal Mather

Acknowledgments

It is impossible to acknowledge all the people who have contributed to my thinking over the years. Clients, consultants, authors, and lecturers—all have helped me to develop my current point of view. It is this gradual development of my thinking that has culminated in this book. Thanks to everyone who directly or indirectly influenced my growth.

Thanks also to my wife, Jean, who gave up several years of vacations to type many revisions and numerous insertions. Her unwavering support helped get this book from concept to reality.

1 | How Effective Logistics Can Make Your Company More Competitive— and More Profitable

Logistics is a term coming into vogue in North America to cover the total flow of materials for a company (from vendor to factory to customer), its associated planning activities, and information flow. The term is already widely used in Europe to cover this spectrum of activities. Until recently, logistics in North America was a military term covering the support of troops in the field.

However, buying materials from vendors, transforming them in a factory into needed products, and delivering these value-added items to customers is every manufacturer's operating mission. This is how they earn profits. But most manufacturers, especially discrete product, batch producers, perform this mission poorly. The major reason is it is not viewed as a continuous process. Departmental relationships are not organized to make this flow smooth and continuous. And performance measurements don't emphasize logistics as a key objective.

Tremendous benefits are possible for the company that improves its logistics performance. These benefits fall into the four major categories of quality, customer satisfaction, inventories, and costs. To achieve the potential improvements will require a team effort from all departments focusing on logistics as the company's key mission.

EVERY MANUFACTURER'S MISSION: GET IT IN, GET IT THROUGH, GET IT OUT

The mission statement for every manufacturer is to: "Buy the right materials from vendors, process them effectively in the plant, and deliver quality products to customers when needed."

In other words, "Get it in, get it through, get it out!" And this mission must be achieved at an acceptable cost so profits are left over after the sale. As soon as we say, "acceptable cost," most people think about performing this mission efficiently. But the statement does not say efficiently, it says *effectively*. There is quite a difference.

EFFECTIVELY VERSUS EFFICIENTLY: KEY INDICATORS THAT TELL YOU HOW WELL YOU PERFORM

Several indicators can tell if a company is effectively achieving the logistical mission statement. Some key ones follow.

1. *Customer delivery performance.* On-time delivery to your promise, or preferably the customer's request date, is a good measure of effectiveness. Flexibility to changing market conditions, although tough to measure quantitatively, is also a good indicator.

2. *Inventories.* By definition, the lower your inventory, the more effective you are.

3. *Profits.* These are an indirect measure of effectiveness. Market conditions can force prices down so you are unprofitable even though you are logistically effective. The links to profitability will be made later in this and other chapters through the cost reductions effectiveness brings.

4. *Obsolescence.* The amount of obsolete inventories, and their companion, slow moving inventories, tell you a lot about your degree of effectiveness.

5. *Customer complaints.* Effectiveness can be rated by the volume of these, either related to delivery or quality problems. Warranty expense is also a good measure of the quality side of the equation.

6. *Market share growth.* This is another indirect measure of effec-

tiveness. If all the other measures are positive, this factor should also be positive.

These items measure the result of your effectiveness. They do not tell if you are being effective on a day-to-day basis. I'll discuss measures that do routinely check your effectiveness in chapter 11.

Rate Your Own Logistical Effectiveness

I always ask attendees at courses to rate their company's logistical effectiveness against a scale of zero (terrible) to 100 (perfect). I don't ask for a quantitative measure but a qualitative feel of how they think their company stacks up. Responses range from a low of 30 to a high of 70. But all these ratings drop lower later in the courses as we delve deeper into the effectiveness question.

Rate your company now based on your knowledge of how you perform in the key areas mentioned earlier. See whether you will be forced to lower the rating as we discuss logistics in more detail.

My guess is that whatever rating you give yourself now, and what it ends up to be, there will be plenty of room for improvement.

You Don't Systematically Measure Effectiveness

The day-to-day operating performance indexes of a manufacturer focus on efficiency or direct costs. But a focus on these two can push you away from effectiveness. Take a look at how our traditional measurements undercut our effectiveness. For example, how many plant managers reading this can deny they have at times deliberately made things not needed to absorb overheads and create "profits"? And often they did it instead of making products customers needed.

How many supervisors have deliberately run the easy jobs in their area to gain efficiency rather than tackle the needed, but difficult to make, jobs on their schedule?

Consider purchasing people, who scour the world for lower costs. They are rewarded with favorable purchase price variances even though deliveries may be erratic. The result? The factory must now have higher raw material inventories to cover these erratic deliveries. And because of a longer delivery pipeline, flexibility to a changing marketplace often is diminished. You'll find a similar problem in most sales departments. Salespeople are measured on revenues booked, with little consideration to the mix they sell compared to their forecast.

Undue emphasis on efficiency and costs can destroy effectiveness. Some companies have succeeded so well in this direction that they routinely make the lowest cost obsolete inventory in the world!

Which Would You Rather Be, Efficient or Effective?

The answer, of course, is both. But what if you cannot achieve them both? Would you rather be 100 percent efficient and 90 percent effective, or 90 percent efficient and 100 percent effective? The obvious answer is 90 percent efficient, 100 percent effective. Your business results will be far superior.

This suggests that we need to measure and reward effectiveness at least as much, if not more, than efficiency. It's a rare company that does. Read chapter 11 now if you want to get going with a new set of performance drivers.

If we drive for effectiveness, what will happen to efficiency and costs? Will they deteriorate so we become noncompetitive? The answer is "no"; they will also improve. A drive for improved effectiveness will give you lower costs and increased total efficiency. The reverse, a drive for efficiency and lower costs, will not give you effectiveness. The past track record for industry proves this point.

FOUR SPIN-OFF BENEFITS FROM A LOGISTICAL FOCUS

The indicators of logistical effectiveness mentioned earlier define how a logistical focus will benefit a business. The benefits can be broken down into four basic areas that will be directly affected.

1. *Better quality.* Many companies are recognizing the value a quality improvement program can bring. Reducing the total costs of quality, from less scrap and rework to fewer quality people and smaller warranty expense, can reduce a company's total costs considerably.

What is not so keenly recognized is the value of an improved quality perception in the marketplace. Part of the reason for the phenomenal growth in market share of Japanese automobiles, from 0.2 percent of the American market in 1965 to more than 20 percent today, is the perception in the American population that Japanese cars are superior in quality to comparable American or European vehicles. The Hertz rental car study of warranty claims in its fleet of cars, performed in the late 1970s, proved the perception was real. Ford and Chevrolet cars had 7 to 8 times the number of repairs as Toyota in the first 12,000 miles of operation. Consumer Reports routinely rates Japanese cars as having better quality than American cars in their frequency of repair records. American automakers have made dramatic improvements in the quality of their vehicles in the past few years, but how long will it take before Americans change their perception about U.S. built cars and accept them as equal—or better—in quality than Japanese? Perhaps never.

Many of the quality programs now in vogue are not achieving the success that was expected. Statistical process control (SPC), Total Quality Control (TQC), and other programs, although fundamentally correct, lack a driver to force them into a company's conscience. That driver is logistics, a focus on a fast flow of products through a company. Combine a focus on logistics with a good quality program and you have a synergistic attack on the quality problems of the business. The improvement in quality will be fast and dramatic.

2. *Lower inventories.* High inventories are the result of a poor logistics flow. Materials come in, and maybe get through, but they do not get out. This problem holds for all pools of inventory, whether raw materials, work-in-process, or finished goods.

A program that synchronizes the total flow of materials and products to what the marketplace needs will make dramatic inroads into all inventory levels. Hewlett-Packard in Fort Collins, Colorado reports an 82 percent reduction in work-in-process inventories at the same time it increased output 29 percent. A 3M division reports a 90 percent reduction in work-in-process inventories after it focused on product flow. And both companies report that there is still plenty of room for improvement.

3. *Greater customer satisfaction and greater flexibility to changing markets.* Many facets determine whether a customer is satisfied with a product or service. On-time delivery, product availability, after-sales support, and technical performance of the product are only a few.

Not all can be improved by an attack on logistics, but better on-time delivery or product availability quickly spring to mind as being prime targets for a logistics thrust. But what about flexibility to changing market conditions? This may be more important than delivery performance, especially for companies that serve a dynamic and changing marketplace. Improved logistics can increase flexibility to change, mainly through reducing procurement and production lead times.

The ability to be quick-to-market with new products may be of vital importance to a technology- or fashion-driven company. This can be for offensive reasons—to beat the competition to market with a new product—or for defensive reasons, such as when the competition has just announced a new product. It is vital you have an equivalent product in the marketplace quickly to avoid losing too much market share or position. Improved logistics capability will improve your flexibility *and* help you be quick-to-market.

4. *Lower costs.* Our traditional view of business costs is to think of direct materials; direct labor; burden or factory overhead; and support costs such as engineering, sales, accounting, general and administrative, etc. This breakdown of costs suits the accounting system but doesn't show where the potential for cost reduction lies. Cost reduction comes from eliminating all non-value added activities, generally categorized as waste and abbreviated as NVAW.

ATTACK NVAW AND BRING DOWN YOUR COSTS

NVAW stands for non-value added waste. My definition of NVAW is as follows:

> Anything other than the *minimum* amount of equipment, materials, parts, space, information, people and time, which are *absolutely essential* to *add value* to the business.

This is a modified description of waste originally created by Cho and Hay. Mr. F. Cho is a manager in operations at Toyota. Ed Hay is a vice-president at Rath and Strong. Cho's definition of waste appeared in the January–February, 1983 issue and Hay's appeared in the March–April, 1983 issue of *Inventories & Production* magazine.

Examples of waste under this definition are

Handling customer complaints

Warranty costs

Scrap

Material handling

Inspection

Storage

Receiving

"Fix it" engineering changes

Inventories

Many computer and manual reports

Expediting

Shortages

Rework

Machine breakdowns

This list does not even try to identify all NVAW for a business. That list would be far too long. To get yourself thinking in the right direction, see how many others you can add for your business. The key is to realize NVAW exists or is created in every department in the company. And all causes of NVAW can be attacked and reduced.

You may wonder what logistics has to do with these wastes. Can't a waste reduction program attack and reduce wastes, without the guise of logistics? The answer, theoretically, "yes"; in practice, "no."

A focus on logistics or product flow provides the thrust for all the waste reduction activities. It's the key to the Pandora's box hiding all the ills defined as waste. Besides which, it's crucial that all waste reduction efforts improve the logistical process, defined earlier as every manufacturer's mission. A focus on logistics as the rallying point for the NVAW reduction effort will be far more effective than waste reduction by itself.

I ask the people who attend my courses to estimate what percent of the total costs of running the business are NVAW. Answers range from 10 to 30 percent. What's your estimate for your company? Making a company more effective logistically will attack most of these costs for significant bottom-line results.

Interestingly, attacking NVAW attacks the nontraditional cost reduction areas of overheads and support staff. Traditional cost reduction efforts address direct labor and materials, together rarely more than 50 percent of the total costs of the business. And these latter elements have been squeezed down over many years of attention. The big remaining costs, overheads and support staff, have received too little attention. An attack on NVAW will provide far more opportunities for cost reduction than all the traditional cost reduction methods put together.

Ford versus Toyota:
How NVAW Can Make a Dramatic Difference

A survey was done by the Ford motor company of two four-cylinder engine plants, one it owns, another owned by Toyota. These plants make very similar products, have about the same degree of vertical integration, and produce about the same number of engines per year. So, to this extent, these two plants are nearly identical. But look at

FOCUSED FACTORY	Toyota	Ford
Engines/Day/Employee	9	2
Square Feet	300,000	900,000

4.5 x Productivity
2/3 — 1/2 Lower Capital Investment

FIGURE 1-1. Ford Versus Toyota: The NVAW Difference

Figure 1-1 to see to what extent they differ. Productivity differences of 4½ to 1 and capital investment differences between ⅔ and ½ to 1. And both in favor of Toyota.

No, one is not a new plant and the other old. Neither is one fully automated and the other largely manual. You cannot escape the differences so easily.

If you don't believe the productivity figures, cut them in half. Toyota still has a remarkable edge. The main reason: elimination of NVAW. Picture your factory with only value being added. How productive would it be without all the NVAW elements defined earlier, plus others pertinent to your business?

Look again at the space figures in Figure 1-1. A difference of 600,000 square feet, at about $40 per square foot for a fully serviced plant, is $24 million excess investment in the building alone. Why did Ford need this much extra space? To store inventories. How's that for NVAW?

I recently visited a company in Australia that produces cardboard boxes. The company has ten plants and two more on the drawing board. I walked through several of the plants as part of my consulting fact finding. My estimate: 70 percent of the prime factory floor space was simply storing inventory. My question: why build two more plants to store inventory when the ten you have are only 30 percent effective? My recommendation: use the ten you have more effectively, save the investment in new factories, and reduce costs at the same time.

Even the machinery investment at Toyota was less than Ford's. To prove this at this time is too difficult. But read chapter 9 to see how Toyota could make the same number of engines with fewer machines and have more excess capacity left over than Ford to grow the business. This seeming contradiction is not as impossible as it may at first seem.

What does this mean about the selling prices of Ford and Toyota engines? To achieve the same return on investment, Ford must be significantly higher priced. You can't be better enough technologically to offset the price difference. Neither can you market yourself out of this much spread. The only things you can do are (1) lose market share, (2) emulate what Toyota has done, or (3) buy engines from Toyota. I'd choose alternative 2 and try to beat them at their own game. Hence we are back to a focus on logistics.

SUBSTITUTE ENTREPRENEURIAL BENEFITS FOR TRADITIONAL BENEFITS

Many of the benefits of improved logistics are not acceptable in traditional return on investment, internal rate of return, or discounted cash-flow calculations. How do you quantitatively value flexibility to a changing marketplace or the ability to be quick-to-market? How about better on-time delivery or the value of an improved market perception regarding the quality of your products? Most of these benefits are too "soft" to be included in financial justifications.

However, from an entrepreneurial viewpoint, these benefits may be worth far more than the traditional "hard" benefits of improved productivity or reduced direct costs. They may mean the difference between a growing market share or a declining share, or the difference between survival or bankruptcy. You must look at the value of improved logistics from an entrepreneurial perspective, covering the total business. The results will be worth the effort.

A NEEDED CHANGE IN FOCUS

Our traditional method of organizing to improve logistics is shown in Figure 1-2. All departments that are part of the total logistics chain operate to their unique set of objectives. A staff group, probably called production and inventory control or materials management, is tasked to integrate these various departments. Usually this coordinating department is part of the manufacturing sector, and so viewed with some suspicion by the other business activities. Moreover, their position in the organization gives them no authority over design or sales (the best they can do is hope to influence behavior in these areas) and limited control over procurement, manufacturing, and distribution.

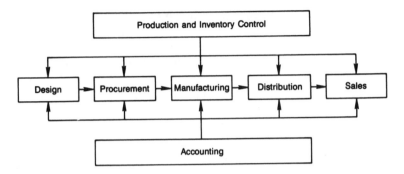

FIGURE 1-2. Traditional Logistics

To top it all off, accounting measures each department to efficiency or cost criteria, not effectiveness. The result is a disintegration of the logistics process. Is it any wonder that people rate their company's logistical effectiveness so low?

What is needed is a radically different view of logistics, diagramed in Figure 1-3. It must become a common focus. I do not mean it must become the only focus. Other criteria are also important for a business. But all departments can help or hinder the logistical process by decisions they make. For example, designers can specify materials and designs that are logistically friendly or unfriendly. Industrial engineers can select machinery and layout plants to improve or impede the flow of products through the plant. Sales can book orders in a smooth or erratic manner. Accounting can set credit and billing terms

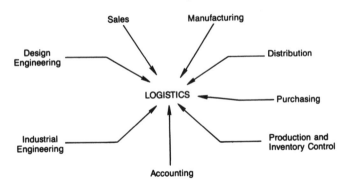

FIGURE 1-3. Industry's Common Focus

that help or hurt a smooth flow of product. All other departments have the opportunity to help or hurt the overall logistics chain.

Only with a common view of the total logistics objectives will each department take those actions or set those policies that benefit the process in total.

2 | The First Step Toward a Bigger Bottom Line: Look at Your Business As a Flow Process

Logistics is a tough process to visualize. Many activities are included, often involving tens of thousands of different parts, raw materials, or finished goods, thousands of vendors and customers, hundreds of factory work centers, and many people in several disciplines. Few people understand how all these data relate and fewer still how all these relationships should fit together to optimize the business.

Logistics is tough to visualize if you focus on the details. The amount of information and different activities that must be performed are overwhelming. But if you step away from the details, the actual process will be seen to be quite simple. Once the overall process is made simple, the details will become simple also.

FIVE BASIC FLOW ELEMENTS

No matter what they make, all companies must perform the same five basic planning and control elements. These five are

1. *Master schedule.* A future product output plan for the plant or business.

2. *Flow rate (capacity) plans.* Calculations of the necessary rates of output all business resources (vendors, factory work centers, and supporting departments) must produce to execute the master schedule.

3. *Flow rate (capacity) controls.* Measures of actual production rates of all resources compared to the calculated plans. Actions must be taken to handle significant deviations.

4. *Sequence (priority) plans.* Calculations of the sequences of activities all business resources must follow to execute the master schedule.

5. *Sequence (priority) controls.* Techniques to ensure the actual sequences are achieved to match the plans.

THE OIL REFINERY ANALOGY:
THINK ABOUT YOUR COMPANY AS A FLOW PROCESS

The best way to visualize these elements is to think of your company as a flow process. I have chosen an oil refinery as my example but could as easily have chosen a brewery or any other flow process. Visualizing your company as a refinery, regardless of your product or process, will make many logistical concepts and techniques easy to grasp.

A group of people operating some specific machines are analogous to a pipe in an oil refinery. Product flows across these machines and out to the next process just as oil flows down a pipe. Vendors deliver a flow of products to your receiving dock just as tankers deliver crude oil to the refinery. The engineers who design products and release a flow of specifications into the factory or to the vendors are just another pipe in the oil refinery. The factory output is a flow of products going to customers or a distribution center just as tank trucks deliver gasoline or fuel oil. All these activities can be viewed as a series of pipes transporting product or information from one place to the next.

THE MASTER SCHEDULE:
THE PIPELINE FROM THE BUSINESS TO THE MARKETPLACE

The master schedule is shown in Figure 2-1 as a pipe leaving the plant or business and entering the marketplace.

The marketplace is shown as an irregular shape to indicate it is a dynamic, indeterminate entity. Customer preferences, a changing

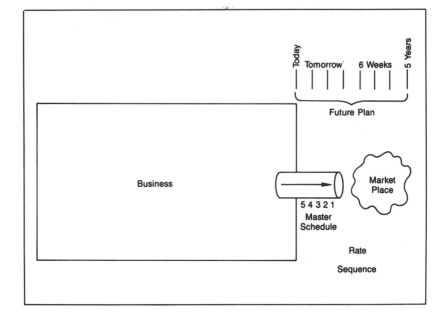

FIGURE 2-1. The Master Schedule Pipe Line

economy, actions of competitors, and so on all cause marketplace dynamics.

The business is shown as a rectangle, suggesting a much more determinate entity. The factory is a certain size, and yes, it can be expanded or contracted, but that takes time. There are a certain number of machines in the plant, and again, these can be changed, but only over time. A certain number of people come to work each day. This number can be changed or the hours worked changed, but again this takes time. There is a fairly well-defined rate of change that is possible or manageable for each of these factors.

No such contraints exist on the desires of your marketplace. The master schedule is therefore a set of numbers that bridges between the indeterminate world of the marketplace to the more determinate world of the business.

The master schedule "pipe" defines the products we want to leave the factory to support the marketplace sometime in the future. It can include deliveries direct to customers, to the distribution center,

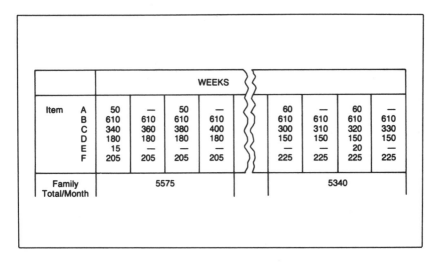

FIGURE 2-2. Master Production Schedule

or to affiliate plants for further processing. Hence, its contents likely include end products, spare or service parts, and unfinished kits of intermediate items. A typical master schedule is shown in Figure 2-2 as a matrix of product, quantity, and time. The products are grouped and summarized into product families for management review. Frequently they are also summarized for gross resource analysis to see if this plan is a feasible one.

In the context of Figure 2-1 the master schedule extends out into the horizon traditionally occupied by the strategic plan, perhaps up to five years in the future. Hence, these numbers encompass long-range planning (what products and resources do we need), intermediate-range planning (what staffing levels, vendors, financial results must we have), and short-range execution (what do we buy, make, and sell today).

This is not the classical view of a master schedule's horizon. It is normally reserved for a shorter time span than the strategic plan. Other terms, such as production plan or master plan, are used to complete the total planning horizon. Please allow me this license to avoid adding complexity to the picture.

The master schedule gives us two critical pieces of data. One, what rate of output do you plan to achieve in the future? This could be expressed as dollars/month, units/week, tons/shift, and so forth. The diameter of the pipe in Figure 2-1 represents rate of flow.

The master schedule also defines the sequence of output—what products do we want to ship first, second, third, and so on. This will probably be based on customer delivery promises, warehouse needs, or other demand criteria. The numbers 5, 4, 3, 2, and 1 in Figure 2-1 depict sequence. It's obvious that sequence concerns run over a shorter horizon than output rate concerns. Sequence changes can be accommodated over a shorter horizon than volume changes in almost every industry.

FLOW RATE PLANNING:
CAN THE "PIPES" SUPPORT YOUR OUTPUT?

Flow rate planning is a clearer term than the more correct technical term, "capacity planning," because of confusion with the word *capacity*. My definition of flow rate is what you want or can reasonably expect to flow through a work center today, tomorrow, and in the future. This is obviously a composite problem of machinery, people, materials, money, and Murphy that ever-present perpetrator of Murphy's law. All together they give you a measure of your expected output from a resource at some point in time.

Flow rate planning, shown in Figure 2-3, starts from the master schedule's planned rate of output. All factory resources, whether finishing work centers (for example, final assembly and test), intermediate work centers (for example, subassembly production), or primary work centers (for example, component fabrication), must be checked to see if they can flow production at the rate necessary to execute the master schedule. All other resources along the product flow axis—vendors, for example, and support groups, such as engineering and drafting—must also be flow-rate tested against the master schedule's desired rate of output.

To return to the oil refinery example, this process checks to see if the "pipes" in the business can flow products or information at the rate needed to support the output plan. Any pipe that might flow too much or too little must be identified and fixed. A pipe that flows too much will mean excess unusable inventories. A pipe that flows too little will mean shortages, a failed master schedule, and the attendant consequences of failing to support the marketplace. It will also mean

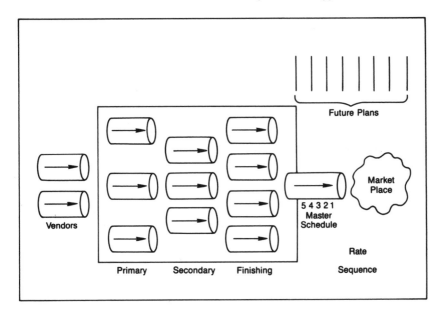

FIGURE 2-3. Planning the Upstream Pipes

excess inventories from all the other pipes producing at the desired rate while shipments are constrained by the small pipe.

The Critical Difference Between Desired Flow Rates and Flow Rate Capability

It is important to separate clearly in your mind the difference between what a work center (pipe) is capable of flowing and what we want it to flow at a point in time. A work center may be capable of flowing twice as much as the master schedule demands. But actually flowing twice as much would not benefit the business. Inventories would grow, cash flow would decrease, and costs would increase because you'd have to look after the excess inventories (NVAW).

Flow rate planning does not pertain directly to what pipes are capable of flowing but to the *needed* flow rates down the pipes. You are after balanced flow rates in all pipes, not the same capability in all pipes to flow products. Hence, if a work center is capable of flowing twice as much as the master schedule needs, we want the pipe to be flowing half full, exactly matched to the master schedule's output rate.

Solving Constrained Flow Rates: Industry's Toughest Job

If the flow through any upstream pipe is less than required to execute the master schedule rate, and this upstream pipe cannot be expanded or supported by flow through other pipes, such as subcontracting, then the master schedule must change to suit the achievable flows of the limiting upstream pipe. The latter step is the most difficult job to do in industry, but it is an essential step if you want to be in control of the business. Overloading a factory work center will not increase the flow rate of products out the door, it will *decrease* it as explained in chapter 7.

Multiplant Considerations

The factory work centers in Figure 2-3 are defined as primary, secondary, and finishing. More or less delineations of the process could be appropriate, depending on the product and degree of vertical integration.

These separate processes could take place in three separate plants. It is not unusual to have several plants that belong to the same company in a product flow line. The output from one becomes the input to the other. But in this case, it is likely that each factory has other customers, not just its affiliate factory. This complicates the analogy a little but does not negate the concept of all these pipes flowing at the same rate, balanced to the end factory's master schedule. It means that the feeder factories must have larger pipes than needed to support the affiliate factories, since the excess flow will go to their other customers.

Understanding the Flow Rate Variables

The real problem with flow rate planning is the number of variables, such as scrap, absenteeism, machine breakdowns, shortages, and rework. These all control a resource's real capability to produce product.

The flow rate variable is shown in Figure 2-4. The outer circle indicates a cross section of a pipe in the plant. The outside diameter is the pipe's capability to flow product without problems. The circles inside are the problems that diminish this potential capability down to the actual capability. Max Close from AccuRay Corporation says to view these circles as cholesterol in your arteries—a great analogy. (Another term for them, of course, is NVAW. Maybe you are getting

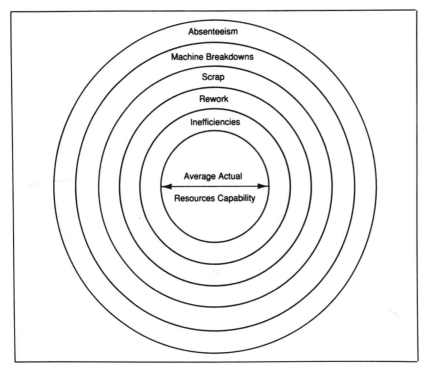

FIGURE 2-4. Defining Average Flow Rate

an inkling of how Toyota can have fewer machines than Ford in the engine plant example of Figure 1-1, make the same number of engines, and still have plenty of capacity left over to grow the business.)

If everyone came to work every day, there were no problems such as scrap or rework, and everyone worked at 100 percent efficiency, then your potential output from the factory would be enormous. But this is not today's real world. These problems and others will occur. Murphy will make sure of it.

All these problems detract from the potential output, and what's left is your average flow rate capability. But the prediction of flow rate capability is made more difficult because these problems vary in intensity from day to day. Who can predict when the next machine will break down and for how long? How much scrap of which items will occur next? How many people will be absent tomorrow?

Because of the unpredictability of these events, the best you can do is plan on averages and react to actuals. This means an aggressive plan to monitor actual flow rates and take action when unexpected deviations occur.

Flow Rate Control

Because there are so many unpredictable variables to consider when planning resource flow rates, it is obvious that actuals must be closely watched to detect significant deviations from plan. This is analogous to flow meters in an oil refinery's pipes, as shown in Figure 2-5. Large flow rate deviations must be identified and a quick solution to the problem found. In extreme cases, the master schedule's desired rate may have to be adjusted temporarily to suit the flow rate capability that is truly possible.

Vendors must be included in this process, as must other supporting resources such as engineering and drafting.

Think of vendors simply as work centers not under your roof or your management. Otherwise they are identical. We treat them differently from our own resources today for no good reason. We must start to schedule and control vendors as if they were another one of

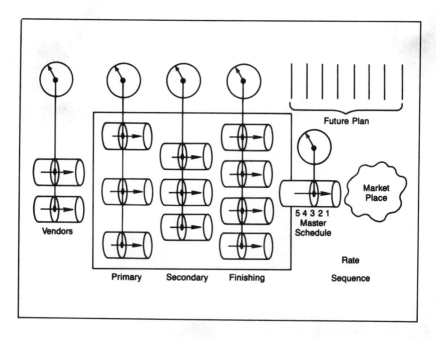

FIGURE 2-5. Controlling the Flow

our factories, a short distance away. Only minor allowances should be made for the change in ownership and lack of your direct management.

The same holds true for support activities such as engineering, drafting, industrial engineering, planning, scheduling, or purchasing. Any group that is directly tied to the flow of products from vendor to customer must be viewed as a pipe. Its flow of product or information must be flow-rate planned and flow-rate controlled to ensure it exactly matches the master schedule output rate.

SEQUENCE PLANNING:
THE RIGHT WORK THROUGH THE RIGHT PIPES
AT THE RIGHT TIME

I have chosen the term "sequence" over the correct technical term "priority" because the word priority often conjures up urgency in people's minds. Urgency is not meant here. What is meant is that the flow of work through a resource must be arranged in sequence with the most needed item first, second most needed item next, and so on.

Naturally, this sequence must support the sequence defined in the master schedule. It's useless to get the flow balanced but have a resource working on the wrong things. High inventories and poor customer service will result from this just as surely as from unbalanced flows. This holds especially true for fabrication and assembly plants, since an assembly department must have matched sets of parts to produce products efficiently. Unmatched parts are useless.

The oil refinery as an analogy breaks down for assembly operations. An oil refinery is a disassembly process, breaking crude oil down into various constituents. A brewery may be a better analogy for assembly plants because it is an assembly process, blending and processing various ingredients, then packaging them into a wide variety of end products. Change the analogy for your company if a brewery makes more sense to you, (and also may pique your interest!).

Out of sequence production is a problem for all industries, not just for fabrication and assembly. It's a critical element that must be planned well.

USING SEQUENCE CONTROL TO RESOLVE PROBLEMS QUICKLY

Like flow rate plans, sequence plans are subject to upset and the ministrations of Murphy. Sequence control disciplines all work centers, including vendors, to produce the right things at the right time.

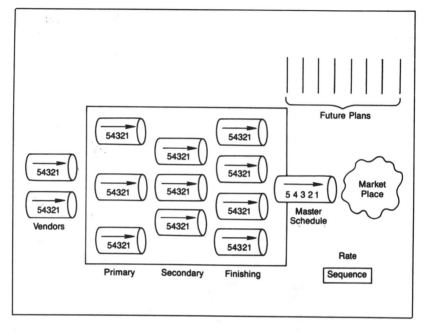

FIGURE 2-6. Sequence Planning and Control

Significant deviations must be resolved quickly to ensure getting a synchronized flow that matches the master schedule's desired sequence. Sequence planning and control are depicted in Figure 2-6 by the numbers 5, 4, 3, 2, 1 in each pipe.

WHY YOU SHOULD AIM FOR UTOPIA

The first thing to do is agree these five basic elements are the key to effective logistics. An excellent output plan called a master schedule that meets the marketplace needs, its output rate supported by balanced flows from all upstream resources, and its output sequence followed by correct sequences in all the same upstream resources, would be perfection. If you want the master schedule's flow rate or sequence to change, then all flow rates and sequences at upstream resources must quickly change to stay in balance. Anything that drives you away from a balanced flow and synchronized sequences will have a negative impact on the business.

Many factors affect your ability to achieve this degree of

perfection—utopia—but it has to be your number one objective. I will suggest utopia several times to you in this book. I do not do this because I think you can achieve it but to make you think what your ultimate objectives should be. With these in mind you can now lay out a plan to move in that direction—and certainly not away from it.

Unbalanced Attacks Are Devastating

You will start to move toward utopia when everyone understands that the five basic elements have equal importance. This is not traditional in American industry. Most people place too great an emphasis on sequence. The majority of people at all levels are concerned with hot lists, shortage meetings, and the latest priority order. All of us, including the general managers reading this book, have been, or still are, great expeditors.

The same emphasis on sequence also appears in operations software. The various software packages (currently about 200, with more coming every week) emphasize their ability to provide sequences, on the minute by the minute if you'd like. Material requirements planning (MRP) and shop floor control (SFC) are largely computerized hot list systems. Little attention is given to flow rates.

What will go wrong in a factory that doesn't know if the flow rates of upstream resources are balanced to the desired master schedule output rate? Chances are they are unbalanced, with some flows producing too much and others too little, but they can change sequences on the minute by the minute. One obvious answer is they have high inventories. The out-of-balance flow will guarantee that.

Another answer is they give poor customer service. You can't possibly get all the things you need through a resource that is not producing enough in total. As a result, shortages will cause late deliveries. Overheads will be high to transmit these changed sequences to all affected work centers and to try to cope with the failures to react. The factory may be very busy but it would be ineffective, building inventories and coping constantly with shortages.

A Balanced Approach Is Crucial

The utopian view given earlier, of an excellent output plan called a master schedule, with its output rate supported by balanced flows from all upstream resources, and its output sequence followed by correct sequences in all the same upstream resources, can come only from a balanced attack on the five basic elements. None can take priority over another.

What would you call a company that has achieved all of the following:

- the flow rates and sequences from its vendors are exactly synchronized to the flow rates and sequences in its primary resources;
- the primary flow rates and sequences are exactly synchronized to the flow rates and sequences through the secondary work centers;
- the secondary flow rates and sequences are exactly synchronized to the flow rates and sequences in the finishing areas;
- the finishing flow rates and sequences are exactly synchronized to the flow rates and sequences defined in the master schedule; and
- the flow rates and sequences in the master schedule are exactly synchronized to the marketplace needs.

The answer: perfection.

What Are the Benefits?

The benefits of a balanced approach will be enormous.

1. *Better customer service.* Our utopian company will execute the master schedule 100 percent. Assuming the master schedule was made well, with a lot of consideration for the marketplace needs, then customer delivery performance or product availability will also be 100 percent.

2. *Low inventories.* These will be at a minimum level, equal only to the processing times of the product. As the processing times of many products can be defined in hours, this means inventory turns of several hundred or even over a thousand.

3. *Short throughput times.* With inventories at a minimum level, the lead time to convert raw materials into end products in the factory will also be short, again in most cases, a few hours. If you extend the ideas of balanced flow rates and sequences to your vendors by considering your vendors as the primary work centers in Figure 2-6, then the lead time from ordering raw materials all the way to delivering products to customers will also be short, a few days at most.

4. *Flexibility to changing markets.* Short throughput times will provide excellent flexibility to a changing marketplace, especially when it comes to sequence changes. If all the pipe flow rates can also

be quickly adjusted, then you have flexibility to both rate and sequence changes in the master schedule.

This idea of being able to change flow rates quickly leads many to believe that means excess capacity, idle a good portion of the time. And that is exactly what it means.

I will prove in chapter 9, however, this does not mean more capacity than you already have. You own enough excess capacity already. The problem is it is not in a form to provide you flexibility today.

5. *Low costs.* Our utopian plant will be a low-cost plant because of the reduction in NVAW costs. Excellent customer delivery performance will eliminate many customer complaints. Almost no inventories means no storekeepers. A balanced flow means little material handling. And our utopian plant will also have reduced or eliminated other manufacturing headaches—poor quality, excess engineering changes, machine breakdowns, and so on.

6. *Effective use of resources.* A resource is being utilized effectively only when it is producing exactly the right amount of exactly the right things customers are buying. Any other condition is ineffective resource use. This company would be using all its resources effectively.

Again, as mentioned earlier, do not confuse effectiveness with efficiency or productivity. You could theorize that these resources may be very inefficient, having high direct and capital costs as a result. And this is possible. However, the things you have to do to make your resources effective will also make them efficient, especially in the broad sense of the word. So, in fact, this utopian company probably has fewer resources (machines, people, and so on) and lower costs by being effective (see Figure 1-1 again) than it would if it were driven to be efficient.

WHAT'S BLOCKING UTOPIA?

An excellent analogy comes from the just-in-time (JIT) movement, shown in Figure 2-7. The boat is sailing on a river, very successfully. Everyone on board is having a great time. But just under the boat are many rocks with very sharp points. If the water falls just a little, the boat will run aground, and the passengers and crew will be swimming for their lives.

Just-in-time is a catchall phrase that defines those actions needed to achieve significant, continuous improvement in business perfor-

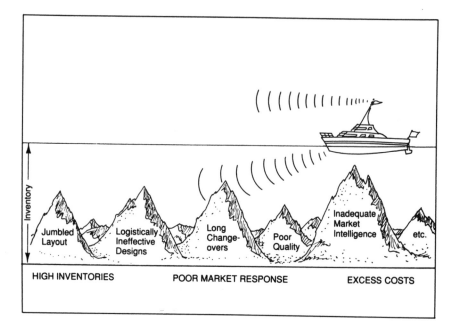

FIGURE 2-7. Just-in-Time Movement

mance by the elimination of waste of time and resources in the total business process. Many people confuse JIT with an inventory reduction program, which it isn't.

The analogy is that the depth of water in the river is inventory, there because of all the rocks (problems). It's the only thing keeping the boat afloat. Hence, large inventories are a symptom of many problems that must be attacked and improved. And, no question, inventories will now reduce, but as a result of eliminating rocks, not as the major objective.

The problems shown in the picture are far from complete. A more complete list is shown in Figure 2-8, but even this is not complete.

Note that most of the listed problems come from the manufacturing sector, where most of today's JIT emphasis is focused. But what about the other business sectors, like sales and marketing, engineering, or accounting? Don't they also have a river and rocks? The answer: a resounding, "Yes." A picture of sales and marketing's river and rocks is given in Figure 2-9, and an expanded list in Figure 2-10.

Long Setups
Poor Design
Functional Layout
Inaccurate Records
Distant Vendors
Restrictive Work Rules
Poor Maintenance
Bad Quality
Slow Decision Making
Poor Market Intelligence
Uncontrolled Changes
Unbalanced Flow Rates

FIGURE 2-8. Inventory Causes—Operations

Logistics—A Total Business Concern

The problems blocking you from reaching a utopian plant are in every department and organization. To grasp the magnitude of the problem, draw the river-and-rocks picture for all business functions. I have given you a start with manufacturing and sales and marketing. Do the

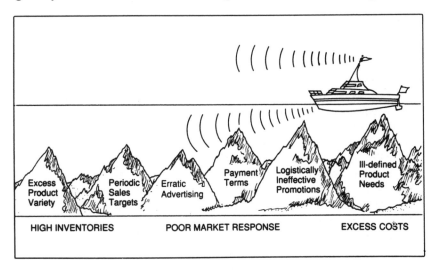

FIGURE 2-9. The River and Rocks Analogy—Sales and Marketing

Periodic Sales Targets
Uncommunicated Promotions
Payment Terms
Erratic Advertising
Logistically Ineffective Promotions
Ill-defined Product Needs
Incomplete Customer Specifications
Excess Product Variety
Poor Discount Structures
Extreme Sales Objectives
Poor or Missing Performance Measures
Created Seasonality

FIGURE 2-10. Inventory Causes—Sales and Marketing

same for accounting and design engineering at a minimum. Then make a list of specific problems in each area, using my two lists to stimulate your thoughts. You'll be surprised at how long each list runs.

You should also realize that the list of problems is a list of NVAW. Can you estimate now what percent of your total costs are in the NVAW category?

Several of the rocks I identified will be discussed in more detail later in the book. I cannot cover them all, so I have chosen those that have almost universal applicability. Use the same methodology I describe to attack your specific rocks.

Inventory Reduction Is Not the Objective

Many people believe the objective of JIT is to reduce inventories. But this is not the objective, as the river-and-rocks picture shows. The objectives are to improve market response and lower costs by removing or reducing the rocks. Inventory reduction is either an incidental benefit or a method of identifying the important rocks.

Why is inventory reduction not the primary objective? Because it's a one-time benefit. When the cash freed up has been spent, no more value can be gained. But improved market response and lower costs are a benefit every day of the week. They are what gives you a competitive edge.

Why Better Systems Alone Are Not the Answer

Some people feel the way to solve manufacturing's logistics problems is with better systems. MRP II is just such an approach. A 20 percent reduction in inventory or more is potentially possible with such systems. (MRP II is a comprehensive planning and control methodology, depicted schematically in Figure 4-1 on page 51. Management decisions are translated through computer software into detailed actions on the factory floor and with vendors.)

But if this is all that is done, then the level of water in the river will drop below the biggest rocks. Instead of being submerged, they will now be islands.

Figure 2-11 is an aerial view of this river. Because of the risk of running aground, we have had to add lots of equipment on the boat, the sole purpose of which is to help skirt the rocks. Travel is still slow and the risk of running aground high. All this equipment is analogous to complex logistics systems trying to keep you out of trouble.

I don't mean to imply that there is no value in these systems. Far from it. Maybe one of your biggest rocks is an information system using invalid techniques or one that is nonintegrated or one too slow for the business conditions. Implementing a better logistics system will dredge this rock out of the river.

But this is the only rock that will disappear. If anything, other rocks will become more firmly entrenched. In fact, systems institutionalize bad habits because many feel that a fast-reacting system

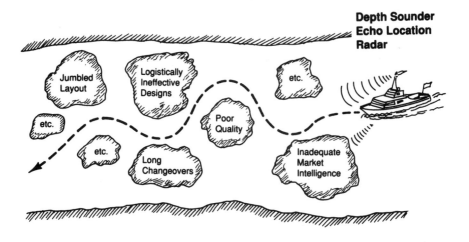

FIGURE 2-11. River Aerial View

helps them cope better with the effects of other rocks, such as machine breakdowns, scrap, and others. Hence, they pay less attention to removing these rocks.

Perhaps the biggest problem with this approach, however, is that we have created a complex system to manage a complex environment. Stop signs, one-way streets, police, and computer-controlled stoplights don't have much lasting impact on the traffic flow downtown. Complex logistics systems don't really help manage industry's complexity, either. The complexity used to managed complexity becomes unmanageable.

CHANGE THE SYSTEM OR CHANGE THE ENVIRONMENT

It's obvious that this is not an either/or condition. Better systems solve informational problems, but they do not solve poor policies, inadequate designs, or complex product flows. Each of these conditions needs its own program for improvement.

This means you need a management process to attack your specific rocks systematically. Removing these flow inhibitors will push you toward a flow process—utopia.

All these rocks are not removable for all businesses. Your specific products, regulatory constraints, volumes, and so forth will determine how far you can go. All businesses can benefit from listing and attacking those rocks that apply to their situation. And with some creative thinking and management pressure, you may be surprised how many rocks can be attacked and ground into pebbles.

3 | How to Tackle Industry's Toughest Planning Dilemma: The P:D Ratio

Most manufacturers have to forecast what their customers will buy. This prediction is used by companies to procure materials and parts and perhaps process them to some intermediate stage before these companies receive a firm customer order.

The first law of forecasting is that forecasts are always wrong. So, often the wrong materials and parts are bought and maybe even processed into the wrong things. How many of you have excess inventories of some things and critical shortages of others? See what I mean? This is the largest amount of NVAW in most companies. Making things no one wants and not making the things they do want means the whole business is actually NVAW! Only a limited number of choices exist to improve this condition. It's critical you understand what these choices are and organize programs to implement your selection.

P:D RATIO DEFINED

This term "P:D ratio" comes from that fantastic but frustrating book by Shigeo Shingo, *Study of Toyota Production System*, Japan Management Association, 1981. It is fantastic because of the concepts it describes, frustrating because the translation from Japanese into English is poor. But it's worth the effort to struggle through it.

Shingo used the term "D:P ratio." I have switched the terms to avoid confusion with data processing, the normal meaning of the letters D.P.

What Is "P"?

"P" is the stacked lead time for a product, from the time that raw materials are ordered until they have been processed through various manufacturing stages into finished goods. This is shown schematically in Figure 3-1. The total P time can be seen as the sum of many individual lead times.

P is also shown pictorially in Figure 3-2 for a company that produces consumer electronics in Brazil. The time elements are expressed in days. There is some variability of lead times at each stage, either because the company is ordering or making different items or because extra pressure is occasionally brought to bear, which speeds things up. The first two stages, buying raw materials and producing components, are done in Europe. The balance is done in Brazil.

As you can see, the total times stack up to almost one year. The other startling statistic is that there are only about three standard hours of labor per finished unit throughout all these processes. We'll come back to this phenomenon later.

Included in this pictorial view is the distribution lead time from the final assembly factory through the warehouse, distributors, and

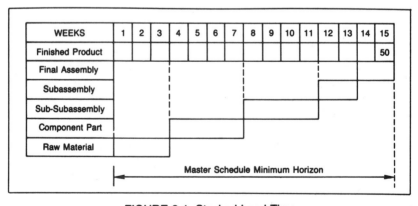

FIGURE 3-1. Stacked Lead Time

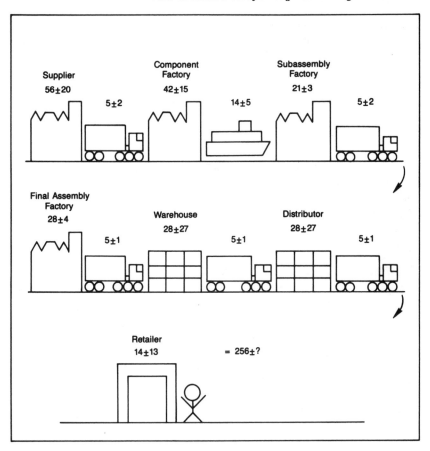

FIGURE 3-2. Lead Time Buildup

the retail stores. The time elements here are the average times that products are stored in inventory plus the variability of these storage times. This is a key point. A large part—in fact the majority—of P time is product storage time, not actual processing time. We'll also come back to this.

I will concentrate on the production and procurement elements of the total P time, because often ownership changes hands from the factory to the warehouse. Hence, distribution time is not of direct concern to the manufacturer. If you are a distributor, I am sure you will be able to take the concepts from the procurement and production world and apply them to the distribution chain.

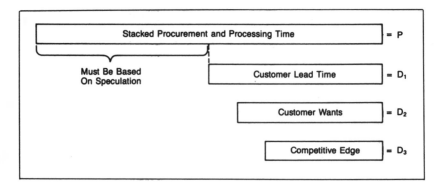

FIGURE 3-3. The P:D Ratio

What Is "D"?

"D" is the length of time from when customers order products until they want them delivered. In other words, D is your customer's lead time.

D depends heavily on the type of business you are in. For a make-to-stock company, it can be very short, a matter of a few hours or days. For assemble- or engineer-to-order, D becomes longer depending on how much of the early portion of the P time can be accomplished based on speculation.

The P:D ratio is shown in Figure 3-3. If P exceeds D, which it does for most of industry, then speculation by the manufacturer is necessary. This speculation is used to buy those materials and perform those processes that must be done before receiving the customer's order. The consumer electronics company depicted in Figure 3-2 has a P time of 256 days with a D time of a matter of hours for the end consumer. The P:D ratio is obviously enormous.

THE THREE Ds

D can be three times. The first, D_1, is what you tell the customer it is. The customer has no say in this decision. The manufacturer quotes deliveries based on standard lead times or based on capacity availability. Most D times are D_1 times.

Second, D can be how long the customer wishes it was, D_2. I have drawn D_2 shorter to suggest many customers would like shorter lead times from their suppliers but cannot get them. This, of course, is not always true.

Third, D can be the length of time that would give you, the manufacturer, if you could quote this time reliably to customers, a competitive edge in the marketplace. I have indicated this as D_3 and drawn it shorter than either D_1 or D_2. This is not necessarily the case either.

I don't mind which of these three Ds you choose to visualize in the P:D ratio discussion. It won't change the basic concepts. But if you choose D_2 or D_3, and they are in fact shorter than your current quotes to customers, D_1, it simply intensifies our discussion. I will use D_1 in my discussion throughout.

THE PLANNING DILEMMA: HOW MUCH ARE YOU AT RISK?

Most manufacturers have a P:D ratio greater than 1 to 1. Some examples are shown in Figure 3-4.

What's your P:D ratio? Is it greater than 1 to 1, as in all these examples? For what length of time (P minus D) are you vulnerable to the questionable accuracy of your speculation? And how far into the future do you have to speculate?

Realize that your forecast starts at the D time from today and extends out to the full P time from today. It's a combination of the P minus D time and where these are in your future horizon that determines your at-risk condition.

Let's take the bearing figures from Figure 3-4. As an example, their forecasts of sales start four to six weeks from today and extend to nine months from today. A huge amount of company cash is spent

Product	P Time	D Time
Consumer Electronics	16 weeks	2 weeks
Bearings	9 months	4 to 6 weeks
Process Control Equipment	15 months	3 to 4 weeks
Furniture	20 weeks	4 weeks
Control Valves	9 months	6 weeks
Military Computers	18 months	6 months

FIGURE 3-4. P:D Ratio Examples

with vendors based on these predictions. A vast amount of company resources are also invested in the product based on these predictions. Now look at your own company and determine your at-risk position based on your specific P:D ratio.

THE INVESTMENT SCENARIO: HOW GOOD IS YOUR RETURN?

Consider what this means from an investment standpoint. A P:D ratio greater than 1 to 1 means you have to invest corporate cash in the hope of getting an adequate return sometime later. P forces you to invest in raw materials, maybe process them into an intermediate stage, or even produce finished goods in the hope of a later sale. You actually make the sale at the D time and transfer from being at risk to being sure of a return.

This risk/return scenario is analogous to a standard financial investment decision, such as investing in stocks or bonds. You are investing at the P time from today with a return at the D time from today. The real question is this: are you getting an adequate return from investing in raw materials and processes—with all the attendant risks of poor forecasts, obsolescence, and damage—that typical financial investments would require? If the answer is "no," then you need to rethink this whole process.

THE DILEMMA DIAGRAMMED

Industry's planning dilemma is shown in Figure 3-5. Forecasts are used to make a master schedule, the link from the marketplace to the business. These numbers are converted into detailed buy-and-make decisions through a bill of material explosion process called material requirements planning (MRP). These buy-and-make decisions apply to all activities in the P minus D time frame.

Intermediate inventories are the result. These inventories could be raw materials, components, subassemblies, or finished goods, depending on your specific P and D times.

Customers' orders or warehouse replenishment orders consume these intermediate inventories and determine when they are processed into specific finished goods. This finishing process can be as simple as pick, pack, and ship for a make-to-stock product. It can be a final assembly schedule for an assemble-to-order product. Or it can involve making or buying specific components or subassemblies, and final assembling them, for an engineer-to-order product.

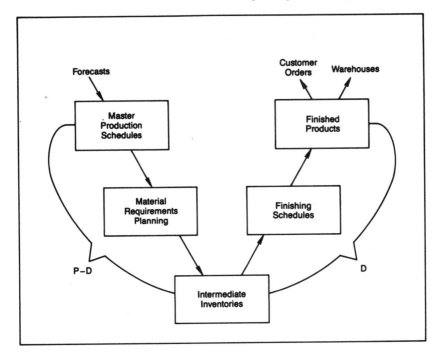

FIGURE 3-5. The Planning Dilemma

Poor Speculation Means You're Stuck with Wrong Inventories

The intermediate inventories are purely transient if the speculation is accurate. The inventories are produced based on the forecast and master schedule and consumed immediately by the customers' orders or warehouse replenishment orders.

If the speculation is faulty, some wrong inventories are produced. These inventories are no longer transient but become more permanent, at the same time as you have shortages of the things the marketplace needs. Can you think of a worse case of NVAW? How does this sound for a good investment scenario?

Three Ways to Tackle the Planning Dilemma

The planning dilemma can be addressed in three basic directions:

1. Make your predictions or forecasts (the left side of the planning dilemma chart, Figure 3-5) closer to reality (the right side of the chart).

2. Accept that the prediction will always contain some element of error. Provide for contingencies or flexibility in the logistics process to correct the error.

3. Build a process or system to recognize errors in your speculation. Then quickly introduce change into the left, forecasting, side of the dilemma chart.

These three choices are given in order of preference. It's obvious that making your predictions closer to reality will give you the best business results. Inventories will be smallest, customer service highest, and costs lowest.

Adding contingency or flexibility to handle forecast error will cost you in excess inventories and/or excess capacity. But at least customer service will be high and costs reasonable. Introducing change can only cost you. Inventories will be high because you won't be able to cancel all wrong purchase orders or production schedules. Customer service will be poor because you won't be able to get all the new things you need, from either your vendors or the factory. And costs will be high because of all the effort needed to change the myriad schedules in a typical company.

Turn Your Thinking Inside Out

Clearly, the first choice—making better plans—is the right way to tackle the planning dilemma. Surprisingly, however, this is not the traditional way we approach the problem. In fact, we usually approach it in reverse. Most emphasis in the past has been on developing fast-reacting systems so plans can be changed easier and quicker. But all change can be categorized as NVAW. On top of this, as we'll see below, change begets change, so that introducing change only spawns a vicious cycle.

Second in order of emphasis has been on contingency planning. Safety stocks, overplanning, and the use of excess capacity to react to errors is common. The least emphasis has been on making plans that don't need to change because they are correct. But that's where the biggest payoff is.

What's Wrong with Introducing Change?

Before we discuss how to make better plans, let's take a look at what's wrong with introducing change. Your operations people are hired to do three things. They spend their time making plans, executing plans, or changing plans. These operations people, usually in overhead or support areas, have a finite amount of capacity just as a machine or direct labor does. While we don't use the word capacity when we talk about overhead or support people, these people come to work for a certain number of hours every day and have the same sorts of disruptions as direct labor, so by definition they have a certain capacity to perform their three activities.

Best business results will be achieved by focusing everyone's attention on outstanding plans followed up by flawless execution. It's obvious that the need to change your plans means you have failed, either in making the plans or executing them.

If this failure occurs, you must steal capacity from these operations people to introduce change throughout the organization. But now there is less capacity to make good plans and to ensure flawless execution. Chances are that plans will be less valid and execution less perfect, meaning more changes will be necessary. Some companies have allowed this change process to usurp the total capacity of their operating people. They are in a constant state of turmoil, continually reacting to the latest change that hits them.

I am *not* saying that all change is bad and should be avoided. If plans turn out wrong, they should be changed and all related activities resynchronized to the new plan. Following a bad plan to avoid changing it would be ridiculous.

I *am* saying, however, that companies must emphasize excellent plans and flawless execution. The full capacity of all operating people should be focused on this objective. Only if this is not successful should change be introduced, and then under tight control, with acceptance of the resulting costs.

In this book, I am only going to cover in detail how to make better plans and add contingency. There is plenty of material already published on how to introduce change, supported by numerous software packages to facilitate the process. Please check the Bibliography if you need assistance in this direction.

FIVE WAYS TO MAKE BETTER PLANS

I have come up with five different ways to make your prediction of reality closer to reality. These are not in any order of preference, neither is any one necessarily *the* solution. Probably all applied to some degree will give you the best results.

1. Reduce P Time

The longer your P time, the more you are at risk and the greater your chance for forecast errors. Conversely, the shorter your P time, the lower your risk.

A utopian P:D ratio would be one where P was less than D. P equal to D is not bad, but it means you must schedule an order the moment it arrives from a customer. When P is less than D, you have some planning time to see how best to schedule the order. This will allow you to better meet your business objectives as well as your customer's delivery dates.

This utopian condition may not be possible, at least over the short term. But don't let this stop you from attacking P and reducing it. Any reduction will reduce your risk and your forecast error, and it's bound to benefit the business.

A P:D ratio comparison—American cars versus Japanese. A fascinating comparison of P:D ratios was made recently between American cars manufactured and sold in America with Toyota cars made and sold in Japan. (I trust you agree this is fair, comparing cars both made and sold in one country with cars made and sold in another.) The almost unbelievable results are shown in Figure 3-6.

The P time is the length of time it takes the logistics chain from ordering the same raw materials in both cases until these raw materials have been processed into a finished automobile. This negates the influence of differing levels of vertical integration between the two examples.

If you find the difference unbelievable, or think that this difference is because vendors stock materials for Toyota, consider a number of facts. First, as Figure 3-2 shows, inventories simply add to P time, they don't subtract. So, vendor stocks cannot account for the difference. This is a difficult concept to grasp, that stocks of materials don't shorten P times but instead lengthen them. Consider vendors that carry four weeks of their finished goods in stock just for you. This

Company	P	D	*Specific Items in Pipeline Based on Speculation*
General Motors	11 weeks	5 weeks	6 weeks' worth
Toyota	3 days	10 days	none

FIGURE 3-6. P:D Ratios—General Motors versus Toyota

means that items flow into their stockrooms on average four weeks before they ship them to you. Hence, vendors must decide what to make for you four weeks sooner than if they had no finished goods. By definition, therefore, the total P time has been increased four weeks.

On top of this, you don't want your vendors to have inventories because the river-and-rocks analogy applies to them too. What you want is to share in the benefits of their increased market responsiveness and lower costs, which can come about only when they reduce inventories.

Second, go back in time to 1926, when Henry Ford reported the P time for his cars in his book, *Today and Tomorrow*. The time it took to unload iron ore at his River Rouge facility, refine it into pig iron, process it into castings, machine it, assemble it, and finally convert it into a finished automobile totaled a mere 41 hours, which included 12 hours transit time from the engine plant to the assembly plant. Not just sometimes, but as part of a regular process. He obviously operated an "oil refinery." Can you visualize the cash flow such a scenario might give?

The doubting Thomases will say the reason was that Henry Ford made any color car you wanted, as long as it was black. And no doubt, his limited variety, high volume process was a great help. But do you honestly believe 11 weeks is necessary just for variety? Perhaps three days is.

A good question to ask is why Henry Ford chose black for his cars. Black is neither a good color for hiding defects, nor a good color to match. Anyone who owns a black car knows how tough it is to keep clean, and how every nick or dent shows up clearly.

The answer is this: black was the fastest air-drying paint. Only by choosing black could he have such a fast cash cycle, from investment in raw materials to collecting cash from the buyer. He did it because it gave him a shorter P time, a critical element in his decision making. His downfall was in dogmatically clinging to one product. He didn't change his offerings to suit the marketplace. The key is to take his ideas and apply them to a variety of products, something Toyota has done very well.

The five benefits of reducing P. If you still find the comparison of Figure 3-6 hard to believe, I promise to prove to you it is correct in chapter 9. In the meantime, recognize that one of the stated objectives of GM's Saturn project is "to manufacture and deliver one of these cars just days after it is ordered." Maybe GM is finally on to something.

For now, ask yourself how Toyota benefits *if* the figures are correct.

1. Their speculation horizon at a detailed level is zero. They still must predict volumes of cars for capacity planning and to orient their vendors to their needs. But this forecast is necessary only at an aggregate level.

2. Their risk of buying or making wrong inventories is zero. Hence they are using their resources very effectively.

3. Their cash cycle is very fast. They probably receive cash from their customers before they pay their vendors.

4. To achieve this fast throughput time, they have created a flow process. This means they must have attacked disturbances causing NVAW, such as quality problems and machine breakdowns. And because they have little or no inventory, they have also reduced the storage and material handling NVAW. So their total costs are lower.

5. To achieve this condition they must have focused on logistics in every department in the company. All the rocks in all the lakes—in sales and marketing, manufacturing, accounting, engineering, and so on—must have been crushed into pebbles. So even more NVAW has been eliminated. Toyota has very effective logistics.

2. Force a Match Between Your Predictions and Reality

Another way to solve the P:D ratio problem is to force your predictions to match reality. You can do this in one of two ways.

1. Make your customers' D time equal your P time. This is commonly done with engineer-to-order products, for example, ships, military hardware, or specialty machine tools. In these cases, though, P time also includes the engineering design time prior to procurement and production. Having D equal to P is an accepted method of operating for these and similar industries.

Extending D times for commercial products is also possible if the marketplace allows it. However, it is a dangerous thing to do, because it does not eliminate speculation, it simply transfers it from the manufacturer to the customer. Extending D times happens at various times in the business cycle. When demand increases, making companies busy, they extend their lead times to their customers. Customers are now forced to give their vendors purchase orders over longer periods of time. Even though the documents are called "orders," they are in fact forecasts. Who is more interested in a good prediction of your future, you or your customers? The answer? You are, of course. But by extending D time you have in essence given control of your future to your customers.

A big order book is always considered better than forecasts of the future. But if these orders are constantly changed by your customers, as they improve their own speculation and refine or alter what it is they want, are you better or worse off? Have you solved the problem or only transferred it? What does it cost you to book erroneous orders because you have pushed your customers too far into their speculation horizon, then make plans based on this erroneous information and finally have to cope with the problems and costs of changing these plans as customers revise their orders? Our earlier discussion on the problems of introducing change should help you think this question through.

I am not saying it is never a good idea to increase D to improve your prediction of reality. If the marketplace allows you to do this successfully, and your customers are not pushed so far into their speculative horizon that they make frequent revisions, then by all means do it.

But beware. Those companies mentioned earlier who traditionally extend D time to meet their P time are beset with the problems extended D times give. The customer's product specifications are often changed, sometimes several times over the life of the project, and so are their requested delivery dates. Sometimes the pace of technology means complete or partial product redesigns.

All of this change has to be absorbed somehow. In many cases it is billed to the customer, especially for military contracts. But a large portion of this wasted effort could be avoided by reducing the P time, hence the D time. Companies that can do that will have a competitive edge because of their quicker response to the marketplace and lower costs because they don't have to contend with so many changes.

2. Sell what you forecasted. Many companies can exert significant influence over the marketplace. These companies can prompt customers to buy different items at different times from what they would otherwise. Promotions, D time differences between products, discount structures, salesperson incentive schemes, product availability, and upgrading are all effective ways of selling what you've forecasted.

One good example comes from the American automobile industry. Automobile companies have recently taken to offering low interest rates on loans and cash rebates for certain car models. This is not out of the kindness of their hearts, but because they forecasted a wrong models mix. The low interest rates attract enough people to buy the overproduced cars rather than the ones they would otherwise buy.

Mail order companies often substitute better products to force a match. So do electronic component suppliers. They ship a better grade product at the lower grade price if they have sold out of the lower cost item.

Restaurants are also outstanding examples of using both product availability and discounting as their way of forcing a match. The special of the day, offered a little cheaper than the regular menu items, moves out food selling below forecast. "We're out of such and such an item" forces the customer away from foods that sold above the forecast. No question, these techniques are also used for reasons other than over- or underselling the forecast. A certain food may be very plentiful, so inexpensive, and offered as a special. Other foods may simply not be available temporarily at the food market and cannot be offered by the restaurant. But these techniques are also very effective to help sell what was forecasted.

Your position in the marketplace will determine to a large degree how effective these or other techniques can be when it comes to selling what was forecasted. Your industry's acceptance of these techniques also carries a lot of weight. Select those ideas that will work for you and use them well to force a match between your predictions and reality.

3. Simplify Your Product Line

The more variety in your product line, the more your predictions will be in error. The narrower your line, the more accurate your predictions. This statement is likely to generate a heated response, especially among salespeople. Often, you have to be a full-line supplier and carry a wide variety to achieve a certain market position. Salespeoples' opinions in this regard, and rightly so, are that it is better to have a wide variety, poor forecasts, and some sales, than a narrow variety, excellent forecasts, and no orders.

If you put emotion behind you, however, it is obvious that reducing product variety will improve your prediction of reality. It's also true that many product lines are encumbered with slow selling products that really don't benefit the business. If pruning these from your product line won't hurt sales of other fast-selling items, by all means do it.

And make this pruning process an ongoing one. Many pressures add products to a company's offerings, from salespeople, customers, and engineers. Few, if any, pressures delete products from the offerings. But dead, dying, or unprofitable items left in your line will hurt you every day.

I will come back to this product proliferation concern in chapter 10. It is a key item, affecting effective logistics, true profitability, and a lot of NVAW. I will show you how to evaluate the benefits and costs of a broad line versus a narrow one and suggest some limited choices for you to consider to improve your current position.

4. Standardize Your Product and Process

This process is quite different from number 3 above. It means to standardize the product and process buildup during the P minus D time. Product variability is OK provided it occurs only during the D time period.

What this means is that designers and process engineers must give you a "mushroom" product, pictured in Figure 3-7. Everything

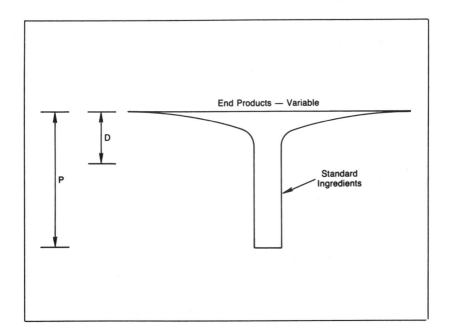

FIGURE 3-7. A Mushroom Design

early in the process is standardized for a wide variety of products. Only during the customers' lead time does the mushroom broaden into product variety.

A product like this helps your prediction of reality because all you need to predict is product families long term. This large aggregation of products will give you a far better forecast to work with.

Converting this forecast into a master schedule and detailed schedules will give you a high probability of having the right intermediate inventories when the customers' orders arrive.

The mushroom product is a logistically friendly product. Compare it to one where there is little or no standardization. Now, you must make forecasts of specific end items long term, at the P period into the future, to plan the nonstandard components. The chance of having the right intermediate inventories when customers' orders arrive is seriously diminished. The business results are obvious.

This idea of creating logistically friendly products will be covered in greater depth in chapter 6. It is a key step in your drive to have effective logistics in your business.

5. Forecast More Accurately

My last suggestion to improve your prediction of reality is to work on the prediction process itself. Many companies don't give this process the attention it deserves. But it is obvious from the P:D ratio discussion that many companies commit huge percentages of their annual expenditures based on forecasts. Effort applied to improve these numbers can often result in significant, fast gains for the business.

Better forecasts result from a well-thought-out, well-controlled process. The actual techniques of forecasting are nowhere near as important as inculcating the process into the business framework.

Keep in mind, however, that the forecasting process runs into the law of diminishing returns at some point. Additional efforts to improve your predictions don't pay off in improved results. When that happens, it is time to stop working to improve forecast accuracy. Instead, you must begin to cope with the inaccuracy that is left.

The whole process of planning and managing the future will be covered in chapters 4, 5, and 6. And as I am sure you have already realized, instituting items 1 to 4 discussed earlier will probably do more to improve forecast accuracy than attacking forecasting itself. Reducing P time, forcing a match, simplifying the product line, and standardizing will all help to improve your prediction of reality. Improving the prediction process itself is just icing on the cake.

TWO WAYS TO ADD CONTINGENCY TO COPE WITH FORECAST ERRORS

I know only two ways to add contingency or flexibility to allow you to cope easily with a poor prediction of reality. These are to build safety stocks or to use overplanning.

1. Safety Stocks

Safety stocks are an amount of inventory of a certain item that will cover errors in your predictions of the future. These amounts can either be calculated mathematically or estimated based on judgment.

The difficulty with this solution is that it involves another prediction problem. For these stocks to be effective, you must first forecast how wrong your forecast of the future is. Then, you have to calculate or estimate the amount of excess inventory needed to handle your forecast of the error in your original forecast.

Does it work? Well, here's what I always ask attendees in my courses: "Who uses safety stocks to handle forecast error?" Most hands go up. "Who has significant backorders with customers?" Most hands go up again.

What does that tell you? It tells me it doesn't work. At least, not as well as many would have you believe. In fact, suggesting safety stock as the solution is an easy out to the tough problems of making the prediction better. Of course, the die-hards will tell you the backorders would have been worse without safety stocks. And having more safety stocks would have solved the problem.

But this all depends on the degree of predictability for safety stocks of specific items. If the predictability is low, you cannot possibly have the right amount of safety stock of each item to keep you out of trouble.

By this I don't mean all the theory and basic thinking behind safety stocks is invalid. Far from it. But it is crucial that you understand its possibilities and limitations. Also understand that, by definition, safety stock is NVAW. Work to eliminate the need for safety stock. You will reduce inventories, improve customer service, and reduce costs simultaneously this way.

2. Overplanning

Overplanning is a technical term that means to add more numbers in the master schedule than you actually plan to produce. These extra numbers, exploded through the bills of material, will put greater demands on inventory and capacity than you really need.

How you react to these numbers will either give you mix flexibility or mix/volume flexibility. If you react only to the excess inventory demands by buying or producing more inventoried items, then you are providing only mix flexibility. This is really nothing more nor less than safety stock.

The only difference with overplanning is the amount of safety stock is controlled through the master schedule. The position of the safety stock, on raw materials, components, subassemblies, or finished goods can be controlled by the time position of the overplan in the master schedule.

If you react to the excess numbers by providing extra materials *and* excess capacity that can be quickly activated, then you have both mix and volume flexibility.

Many companies use overtime to give them volume flexibility. Provided they haven't scheduled overtime to keep up with the original plan and provided they have excess inventory in stock, they can be quickly flexible to the marketplace within the volume limits the overtime can provide. Case studies showing this use of overplanning are described in chapter 7.

How to Use Contingency Planning

Is overplanning NVAW? Of course. And you're up against the same problem that you encountered with safety stock: forecasting the error in your original forecast. Follow my earlier admonitions to improve the plan predictability first. Your business can only benefit by this approach. Use the contingency ideas only as an interim step or to cover the leftover forecast error you cannot eliminate from your business.

WHAT CHOICE TO PICK?

There is certainly no one answer. All these ideas have value at some point in your business' development. And, as mentioned earlier, several are interrelated.

Keep in mind the order of preference stated earlier, however. Improve the plan first, add contingency second, change the plan last. Recognize you are bucking a powerful trend that drives industry to solve the problems in the reverse order. But also recognize my sequence will give you the best business results and certainly the most effective logistics.

4 | Planning Step 1: Create a Strategic Plan That Drives the Total Business

Almost every decision, action, or investment made by a manufacturer is designed to support some future planned event. The P:D ratio discussion in chapter 3, for example, concentrated on specific buy-and-make decisions and actions to support a future sale. Decisions to invest in research and development (R & D) are also designed to support a future product introduction. Building a new plant or adding an extension on an existing one presumes the ability to fill this added space in the future with more productive output, either to support higher future sales of the existing products or to produce a new product. The same goes for decisions to add more or faster machinery or provide new and different process capabilities. Even hiring plans are created to support some future planned increase in output.

This means that the future planning process for a manufacturer must be well thought out, and based on input from those managers and staff best equipped to define what may happen in the future. As predicting the future is a risky business at best, a company needs to routinely review the assumptions it made in coming up with its future plans of action, and correct the course as more current facts become available.

THE PLANNING HIERARCHY BEGINS WITH YOUR STRATEGIC PLAN

A manufacturing company should be controlled through a hierarchy of plans and actions starting with long-range, aggregate business plans, and ending with specific actions to produce or buy something. A variety of plans exist between these extremes, each one differing in level of detail and length of horizon. You must insist on consistency among these plans to avoid conflicting decisions and actions.

Long-range, aggregate plans should directly drive the next level of detail. This next level of detail should, in turn, drive the next level of detail and so on. This means that long-range, aggregate plans are most critical. Errors here will be transmitted into the details of the subsequent plans, and your ability to reverse the errors at that point will be limited.

A common problem is for managers to ignore this hierarchy and allow plans at various levels and for various business disciplines to be developed in isolation from one another. Confusion and chaos are inevitable, as is conflict between the various disciplines.

As an example, the manufacturing division of a company invested heavily in automated equipment to become a low cost, limited variety, high volume producer. At the same time, the sales division structured its selling program to go after specialty accounts. That meant high variety, low volume, and flexibility to change. The conflict is obvious.

A flow diagram of the correct business planning hierarchy is shown in Figure 4-1. It shows the general flow process and indicates planning horizons and time increments. These latter two are meant as a guide only, not the horizons and times you should actually use in your own business. Yours will depend on your products and processes and in some unusual cases could be quite different from these estimates.

TURNING YOUR STRATEGY INTO FOUR ACTION PLANS

The strategic business plan is the key driver of four subplans.

1. *Research and development plans.* This set of plans must define what new products, major improvements to existing products, and additions to existing product lines are needed. That portion of the plan concerning new items is created largely in conjunction with sales and marketing. That portion involving major revisions to existing product lines is most often done in coordination with manufacturing.

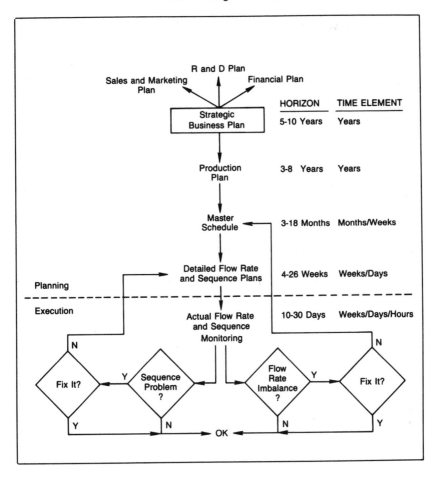

FIGURE 4-1. Business Planning Hierarchy

Cost reductions, ease of manufacture, and technical enhancements are the major objectives of product revisions.

   2. Sales and marketing plans. These action plans concern themselves with pricing strategy and method of sales, that is, through retailers, wholesalers, jobbers, an internal sales force, and so forth. Advertising and merchandising programs, the choice between a full line or a specialty house, and price and quality position in the marketplace must all be considered and defined. It's obvious that many of these decisions can be done only in conjunction with the R & D and manufacturing decisions.

3. Financial plans. Few companies can generate their total needs for money through internal cash flow alone, so some form of financing is often required. Both your timing and your mix of debt and equity financing can have a significant impact on achieving your business goals.

You can only estimate future cash needs by collaborating with the other three groups, R & D, manufacturing, and sales and marketing. Only with their help can you predict future inventory levels, accounts receivable, and capital investments—all of which are vital to calculating a cash needs analysis for a business.

4. Production plans. The production plan concerns itself with resources, both physical and human, in the manufacturing sector. I have only shown in Figure 4-1 the complete flow of information for this one plan because it is the one directly related to logistics and operations. However, it is obvious that this is not a stand-alone plan. The production plan needs inputs from the other three subplans before it can be created successfully. We'll discuss the production plan in more detail later in the chapter, and we'll follow the flow of information from the strategic business plan all the way to specific actions taken to buy or make something.

These four plans are consistent because they are all linked directly to the strategic business plan, the starting point.

Look carefully at the lines and arrows between the various activities. They show a dependancy relationship that must truly be present. Create these links to ensure that senior management's strategic decisions are actually implemented as discrete actions in the business.

HOW THE STRATEGIC BUSINESS PLAN DRIVES THE BUSINESS

A strategic business plan sets the foundation for all subsequent decisions for the business. It defines the long-term business goals, which may be to achieve a given return on investment or a certain share price, gain a defined level of market share. These goals can change over the strategic planning horizon. For example, market share could be your goal only for the early horizon of the plan. Then, you might switch to achieving a given return on investment second and the share price last.

What the Plan Should Cover

The strategic business plan addresses the major variables for the business. It considers and defines such things as: the types of products to be offered; the sales and marketing process; the scoped marketplace (for example, domestic only or including selected areas of the world); manufacturing and sourcing locations (again, domestic or overseas); warehousing and distribution alternatives; and financing needs and methods.

The strategic planning process starts with a collection of facts about the business, its strengths and weaknesses. This information is usually organized by broad product family and includes sales forecasts for these families over broad time intervals. The plan also includes an assessment of major competitors' strengths and weaknesses to find areas of opportunity that exploit the competitions' weak points.

In addition, the plan also forecasts future actions by competitors and impending changes to the current environment, such as wage increases, raw material supply availability and cost, and potential price increases for the product. By performing a sensitivity analysis on a pro-forma profit and loss statement and balance sheet, you can show which events are likely to make achieving the business goals difficult. Using that information, you can plan specific actions to counter these negatives. Redesigning products, reducing labor or material content, investing in capacity to gain market share, attacking new market segments, and using new sales channels are just a few of the actions you can take.

How Confusing Plans Create Production Chaos

The strategic business plan is the most important plan for the business. It establishes the strategy to be followed by the four main disciplines: sales and marketing, research and development, finance, and production. Their strategies in turn are to achieve the business goals, both long and short range.

The full implications of a given business strategy must be understood and each of the four main disciplines must construct their subplans to support the overall strategy. Unfortunately, this is often not the case, creating the conflicts mentioned earlier.

What happens when the plans don't mesh. A company I visited makes products with a high engineering content; a large portion of their business has to be uniquely engineered to each customer's order. So engineering, an overhead department, is a work center, just like any production resource, with a need for capacity that supports current and anticipated customer orders. This company's detailed strategic objectives were as follow:

- Gain market share—grab opportunity business.
- Keep the work force stable—no hiring and firing.
- Promise quick deliveries—never lose an order based on delivery.
- Reliably meet customer promise dates.
- Have low inventories.
- Meet the profit plan—no hiring of indirect or overhead people allowed.

These objectives are motherhood terms and could come straight from any textbook on business management. Without a well worked out set of subplans to achieve these goals, however, they are in direct conflict with one another. For example, grabbing market share means shipping more products. You don't gain market share by booking orders but by producing and shipping them. How can you do this with a stable work force?

Overtime gives some flexibility but this company was working on average 22 percent overtime. Just to produce the orders already booked, some departments were working as high as 30 percent. You can't expect good products to be made efficiently when people are working 48 to 52 hours per week consistently. Absenteeism is bound to rise, quality is bound to suffer, and efficiency is certain to drop. The net effect will be an output of good products only slightly higher than would be produced during a standard 40 hour workweek. And since the company is working this much overtime just to keep up with orders booked, what is left for additional opportunity business? The answer: zero.

Subcontracting is a possible way out of this box. But this means having subcontractors who can make your products to your specifications and quality levels, and who are prepared to increase their flow rates quickly when your business increases as well as keep up with the demands of their other customers. One company strategy, then, could be to cultivate a cadre of subcontractors willing to react like this. It may mean giving them some work all the time, even when your own plant isn't busy, to get this surge capability when you really need it. Or you may have to pay a retainer to get the quick use of capacity when necessary. Unfortunately, the company in question did not have either tactic in place.

Promising quick deliveries and being an on-time supplier with low inventories are also two conflicting objectives unless there is a strategy to make them mesh. To do both, you need to be able to get unique materials or purchased items quickly or have a product design that contains many standard items. Hence, either a vendor or an engineering strategy must be in place. Again, in the case of this company, neither were.

Moreover, gaining market share when the primary resource, in this case engineering, has a frozen work force level, is ludicrous. A profit plan and its associated budgets are a guide to be broken when business conditions warrant. How can you accept this kind of straitjacket when the business conditions are dynamic?

Your Tactics Must Depend on Your Strategy

As you can see, key tactical plans must be developed in concert by each business discipline to achieve the strategic objectives. These plans must grow out of the strategic business planning process. They also must be logical with reference to all disciplines and the business goals.

Wickham Skinner, in his book, *Manufacturing in the Corporate Strategy*, (John Wiley & Sons, 1978) discussed the conflicts that often exist in strategic business plans. He suggests that you can't merge highly varying strategies. A high volume, limited product line producer cannot also be a low volume, varied products producer that can change quickly to take advantage of a dynamic marketplace. Every business—or product line within a business—has to be positioned within the triangle of volume versus variety at the point that best suits its business goals, as shown in Figure 4-2. A medium variety producer can have medium volumes (shown by the dashed lines) but cannot have high volumes that fall outside the triangle. Similarly, a medium volume producer cannot have high variety.

While Skinner's argument has merit, I don't support it completely. I believe you can be relatively high volume with a wide variety of products and be flexible too. The tactical plans to achieve this are more complex and must be much better defined. This will push the envelope of volume and variety wherein a company must operate to look more like that shown in Figure 4-3. Higher volume and higher variety are compatible, as shown by the change in the dashed lines.

Many of the recent improvements to managing the operations side of industry have made this envelope larger. Integrating all functional activities so their individual actions support each other, instead

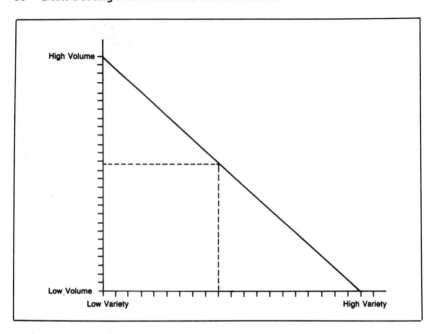

FIGURE 4-2. The Volume Variety Spectrum

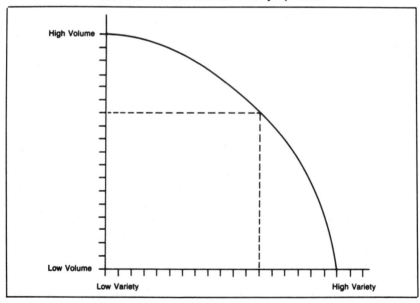

FIGURE 4-3. Enlarged Volume Variety Spectrum

of being in conflict, is one of the driving forces behind our changing view of manufacturing. By eliminating the rocks in the lake (see Figures 2-7 and 2-9), you can also blur the distinction between high volume/low variety/low cost and low volume/high variety/high cost production. As we get closer to the automated factory of tomorrow, further advancements will push this envelope farther still—one of the key objectives of the automated factory.

The strategic business plan sets the stage for all subsequent operations' activities. It's critical that your competitive strategy for your business and products be well thought out and that each major discipline derives subplans to support this strategy. Conflicting plans— such as letting manufacturing invest in specialized machines and tooling to make long runs of limited products cost effectively when your sales department wants to compete with small lot, customized production—are the sure road to disaster.

Once-a-Year Planning Is Not Enough

Strategic business plans are not static. The typical process of reviewing and creating business plans once per year is probably not adequate for the majority of businesses today.

Annual business planning suited a more leisurely business climate. It is a rare company that can take this approach and be successful in today's environment.

The key people involved in the strategic planning process are the heads of each discipline plus the general manager. Other middle managers will support their activities, but the hard work should be done by the top level people. Only when they agree to a set of business goals and develop synergistic plans to achieve them can you really say, "We have a strategic business plan."

HOW THE PRODUCTION PLAN WORKS

The production plan is the operations subplan that supports the strategic business plan. Its purpose is to help you evaluate the resources needed to execute the business plan far enough ahead of time to get those resources in place when needed. The evaluation has to include both the type of resources (machines, buildings, people, and vendors) as well as the quantity of resources.

Business Strategy Must Be Considered

A strategy that requires low-cost production of a few products in high volume differs from a strategy aimed at making low volumes of a wide variety of products. Generally, you will need different types of equipment, different levels of skills, and you may even have to locate your manufacturing operations in different countries. A strategy that requires both high volume and high variety will be different again.

The degree of emphasis on market share in the strategic business plan will determine how much excess resources over and above the plan to provide. All plans are wrong to some degree, the error being a function of how far out they are made and the market dynamics you work in. Recognizing this and planning to counter with excess capacity is a critical part of the production planning job. Failure to do so will inhibit production if the market opportunities do materialize, and you will miss your objective of increased market share as a result.

The Link Between Production and Finance

Here is where the production plan must link to the financial plan. As soon as the additional resources necessary to execute the strategic business plan are defined, their costs and date of installation can provide a time-phased view of the cash needed to provide them.

You must test your cash needs against the financial resources of the business to see if your production plan is feasible. If not, alternative ways of providing for production must be explored. These might include subcontracting, as mentioned earlier, modifying or building machinery in-house, and increasing the number of shifts. Reducing inventories could also be a way of unlocking some financial resources for productive needs, but this will take a strategy all its own to achieve.

You May Have to Modify the Strategic Plan

If none of these production plans is valid, the last resort is to go back and modify the strategic business plan. All the arrows in the flow chart from the strategic business plan down to detailed flow rate and sequence plans are double-headed in real life. I chose to make them unidirectional to show the major relationships and the desired direction of information flow. But the real process is an iterative one, not completely hierarchical.

Modifying the strategic business plan could mean you must also modify the other subplans in sales and marketing, research and development, and finance. For example, there's no point aggressively trying to increase market share if the production facilities cannot support the increased sales. It would be better to be more selective in the marketplace and book higher margin orders.

Again, however, I find this uncommon in practice. For example, one of my clients had a plant that was severely overloaded with customer orders. They were working overtime, hiring and training people as fast as they could, and subcontracting to sources who could make their products. It was still not enough. Orders were coming in well above the shipping rate. They had more machines on order, but the lead time on these machines was 12 to 18 months. They were blocked. What to do? Obviously rebalance supply and demand somehow. Since they couldn't increase supply fast enough, the only choice left was to reduce demand.

An analysis of the sales orders showed that the Sales and Marketing people were aggressively discounting their list prices based on the quantity ordered. This was a deliberate strategy to get people to order large quantities and thereby increase market share. But as I said earlier, you only increase market share when you ship, not when you book orders. My suggestion was to make the discount structure less aggressive and let the price changes dampen demand. As a result, the company would get higher net income from its shipments, some of the customers would go elsewhere, and the company's ROI would increase.

I know this is not an easy decision. There are other considerations, such as long-term relationships with customers and the like. But on the other hand, taking low margin orders and not shipping them didn't make much sense either. Something had to be done to get each department to work in concert.

The real problem was the company's managers were not comparing actual events to their plans often enough and were not updating their strategy to suit. As a matter of fact, the general manager was quite surprised at the depth of discounting and couldn't understand why it hadn't been reduced already. The reason: each function was busy looking after its own unique objectives. No one was considering the business in total.

Timely measurement of operating, not financial plans and speedy reaction to the actuals, will be a recurring theme throughout this book, a goal few of us live up to. But it is an essential change we must make in the way we manage if we want to survive.

MATHER'S LAW OF PRECISE INACCURACY

The product definition and time elements used in the production plan can often be treated the same as those in the strategic plan, that is, by large product groupings over broad time periods. Frequently these are good enough for defining resources needed. Sometimes, though, the strategic plan needs to be broken down into more detail to meet production planning's needs.

Beware of this trap, however. When you're looking ahead three to eight years, your estimates of sales will be approximate at best, even though forecasting large product groupings. Any attempt to get a more precise view of sales three to eight years out will simply aggravate the inherent inaccuracy in the numbers. Using this inaccurate data to plan specific resources will mean providing all the wrong resources, even though they have been precisely calculated.

The conflict between accuracy and precision occurs often in industry. We want to be precise but cannot provide sufficiently accurate, detailed data, so the answers we get are in fact inaccurate. Many idle or underloaded machines, not to mention a few factories, are the sad byproducts of companies which lost sight of accuracy in their search for precision.

Here's a case in point. It can take up to three years to get a new machine justified, approved by the capital appropriations review board, ordered, received, and installed. Assuming an amortization life of three years (low in some cases, high in others), you must forecast the production coming off a new piece of equipment three years from today and ending six years from today. Faced with this scenario, you had better go for accuracy in the form of high levels of product aggregation, flexible machines to take care of some error in the forecast, and some extra contingency to take care of opportunity business not forecast.

THE MASTER SCHEDULE: YOUR PLAN FOR GETTING OUT THE PRODUCT

The master schedule is a more detailed view of the future than is the production plan. It is more detailed in terms of both product and time. This amount of detail is necessary because the high level of aggregation in the strategic business and production plans cannot be converted into specific production or buying actions. These specific actions are connected to making unique finished goods or small product families, and they are consequently more time sensitive than the broad time intervals in the production plan. You have to know what to buy for production in a given week or day, not sometime in a quarter.

The master schedule's job is to provide a future plan for the output of the plant in enough detail that detailed flow rate and sequence plans *can* be developed. This doesn't mean that the level of detail in the master schedule should be constant throughout its horizon. Far from it. As we'll see in chapter 5, the master schedule will have its own series of time zones with accompanying levels of detail to suit short-, intermediate- and long-range planning horizons.

The critical point, though, is the master schedule should match the strategic business and production plans in total. By match, I don't mean that every time period should coincide exactly, but these plans should mesh within an acceptable tolerance over some reasonable time frame.

Chaos Reigns When Short- and Long-range Plans Don't Coincide

This is another weak link in many companies. The long-range business plans and the short-range execution plans are often in serious conflict. Here's a good example.

Officials of a company I visited were very concerned about their poor financial results. They were well below their goals for ROI. The company officials asked me to review their detailed planning system and the solutions that data processing had proposed to remedy their problems.

I visited them just after they had finished their annual strategic planning process. Of course, I asked to see the plan.

It was a full, three-inch, loose-leaf binder, with strengths and weaknesses of their company and products, as well as those of the competition, clearly defined. It also specified which products and markets they should address, which ones to retreat from, and why. It was one of the best strategic business plans I had ever seen.

Next, I asked to see the master schedule. My first shock came when the vice-president of business planning asked, "What's that?" I explained that the master schedule provided the detailed interface between the strategic business plan, the production plan, and the actual production of specifics in the factory. He said, "I've never seen one of those. Do you think we have one?"

By definition, everyone has a master schedule. The only question is, at which management level was it created? By telephoning around we finally found someone in the materials department with a document titled "Master Schedule." This schedule drove the logistics system that in turn calculated what specific things should be made or bought.

We quickly compared the strategic business plan numbers with a summary of the master schedule. Not surprisingly, these two very separate plans didn't match. The master schedule covered only the traditional items the company had been making and selling for years. Nowhere was there evidence of the company's strategy for moving from traditional markets and products into the new areas that would give them their required ROI in the future. And it would be pointless to change the detailed planning system when this fundamental link between strategic plan and master schedule was broken.

To strengthen that link in your company, make sure that you build this phrase right into the job description of the person responsible for the master schedule:

> Convert the first X months of our strategic business plan into the details necessary to execute it. Notify key managers if the short-range details deviate from the strategic business plan by plus or minus Y percent for review and decisions.

Figure 2-2 shows that the individual items in the master schedule have been totaled for a given product family. This total must match the strategic business plan and production plan within a reasonable tolerance. If it doesn't, the master schedule, strategic business plan, or production plan, and all associated plans, must be changed to make them conform with one another.

USING ROUGH-CUT CAPACITY PLANNING TO CORRECT FLOW IMBALANCES

Using the master schedule, you can determine the flow rates needed in the upstream workcenters to support your planned output. These flow rates still cannot be defined in complete detail; further definition is required to do this. What you can do is to make an approximation somewhere between the rough plant and machinery estimates in the production plan and the specific flow rate calculations in your detailed planning and scheduling system.

At this point, you have calculated the approximate workcenter flow rates necessary to support the master schedule's flow rate and specific mix of products. Now, compare your required flow rates to what can actually be accomplished in these work centers. Any conflicts must be resolved, either by changes to the resources or to the master schedule.

This activity is technically called "rough-cut capacity planning" but following our modification of terms, let's call it "approximate flow rate planning." This is an essential step to make sure that any gross imbalances between the desired flow rate in the master schedule and the capabilities of support resources are identified and resolved. There is no point in moving to a detailed analysis, with all its complications, when approximations can pinpoint the major problems. These problems must be resolved at this level of aggregation before you move on to detailed analysis, which should be viewed simply as a fine tuning of the approximate analysis done earlier.

HOW TO DEVELOP DETAILED FLOW RATE AND SEQUENCE PLANS

The master schedule, although more detailed than the strategic business or production plans, is still at an aggregate level. The aggregation in this case, though, is in the product make up.

The term "master" means that this schedule usually occurs at only one level in the bill of material. This level could be finished goods, major assemblies or intermediates, or at some specially constructed position to improve the future planning accuracy. "Master" means all detailed schedules will be derived from these numbers.

It also means that there is a bill of material underneath this master schedule which defines all the detailed parts, raw materials, and subassemblies needed to make the items listed in the master schedule. In addition, the manufactured parts, subassemblies, and finished products all have a routing or process sheet that defines which work centers and how many hours per unit of product are needed to produce them.

With this detailed information, it is relatively easy to see how you can calculate what items are needed, compare them to what you have now, figure out what is missing, and determine when the missing items must be available. This gives you the detailed sequence plans for all resources, including vendors.

You can calculate required flow rates in detail by extending the information about what you need and when, by the routing details for all manufactured items. Your flow rates should be tied to whatever degree of detail is in the routing, for example, by work center, piece of tooling, skill groups, and so forth.

The volume of information often found in industry means these calculations generally have to be performed using a computer. This is not just to create the information once, but to keep it current as conditions change. For example, a change in the master schedule means you must quickly calculate what changes must be made to flow rates and sequences to resynchronize the planning. The same is true for other upsets, such as unexpected scrap losses, machine break-downs, engineering changes, and late vendors.

YOUR PLANS ARE USELESS WITHOUT ACTUAL FLOW RATE AND SEQUENCE MONITORING

No plan is worth making if you don't intend to measure actuals against the plan. Plans by themselves are useless. As mentioned in chapter 2, many variables can affect the operations side of a manu-facturing company. This means that timely measurement and quick recovery plans are needed to ameliorate significant problems.

How to Respond to Flow Rate Problems

The worst problem you can face is an actual flow rate imbalance. The smooth flowing "refinery" is now upset, which means you are either building inventories or shortages. The master schedule may even be in jeopardy, putting at risk customer service and the expected finan-cial results.

A flow rate problem must be identified and solved quickly. If the flow rate through a resource is insufficient, the cause must be iden-tified and resolved. If you have an unusually large number of machine breakdowns in a given area, for example, you must off-load to other machines or work overtime in this area after the machines are fixed. Similar approaches must be taken for unusually large amounts of scrap, absenteeism, indirect work, poor efficiency, or late deliveries from vendors. The key point to remember is this: *flow rate lost is lost forever.* You can't go back and reclaim it. All you can do is add more flow in the future to make up for what was lost.

Let's look at the opposite extreme. If the flow rate is excessive, your problem now is to reduce the flow. This means reducing the number of shifts, or reallocating the work force to other activities, for example machine maintenance, training, or quality circles.

Now let's look at still another situation. Let's say you can't or don't want to solve a flow-rate problem. In this case, you have no choice but to change the master schedule's desired rate of flow and rebalance all resources to it. This may mean constraining the demand in some way if now the master schedule is below what is needed in the marketplace, or building inventories, or working ahead on some orders if the master schedule's output rate is now excessive.

What to Do About Sequence Problems

A sequence problem can come about in two ways. First, it should be obvious that if you ignore an insufficient flow rate problem you'll also get a sequence problem. If a resource cannot produce enough in total to support the master schedule, it cannot possibly make all the right things at the right time. Some are bound to be late.

A sequence problem can also occur if the *average* flow rates through all resources are sufficient to execute the master schedule but demands on a specific resource at a particular point in time are too great or too little. Talk to your vendors and supervisors. They will tell you they live with either feast or famine; their resources are either overloaded or underloaded. This is a phenomenon we'll discuss later and define what causes it. It's obvious, though, that some of this irregularity of demand can come from the specific mix of products sold and the specific mix of inventories in the factory. If these are biased toward certain products, then the mix of work required on the work centers will be uneven.

Cross-trained People Are Your Greatest Asset

One solution to flow rate and sequence problems is to cross-train your people in multiple skills and quickly move them from the underloaded resources to the overloaded ones. This means you must pursue a manufacturing strategy to invest in direct labor. In return, you'll get smooth running operations. Union contracts must also be written to facilitate moving people quickly between resources. Reducing the number of labor grades, although superficially more expensive, can be an excellent way of getting labor flexibility and hence better operating results.

Several Japanese companies use an interesting system to encourage labor flexibility. They post charts of employees' names and skills in each area. As each employee becomes proficient in a certain skill, the chart is marked to signify that employee is certified to perform that job. This creates considerable peer pressure to be the one with the most skills and has a direct bearing on who is chosen for better jobs.

At the other extreme, please don't permit a wasteful practice I once saw in a factory in Europe. There, an expensive machine was allowed to remain idle because the operator was absent. No one else was permitted to run it. Several days worth of output from this machine were kept in inventory to buffer the downstream operations when this operator was not there. Worse still, a large pile of work was growing ahead of the machine from the upstream operations, waiting for this operator to return.

Balancing the Load to Solve Sequence Problems

Another solution to the sequence problem is to off-load the work from the overloaded resources and route it to the underloaded ones. This may be more "expensive" in accounting terms; the accounting department would rather you didn't make it at all or made it late than make it at a higher theoretical cost. But making it, even though at "higher cost," will give better operating results than having shortages. Keep in mind too that many of the "variances" accounting will report are fixed cost differences, not variances at all, so there is really no higher cost.

Another way to balance the load is selective overtime for the overloaded areas or, alternatively, selective subcontracting.

Still another choice is to modify some of the basic parameters in your detailed planning system. Batch quantities, safety or buffer stocks, average lead times, bill of material levels, and routing sequences are all flexible to some degree, even though we assume they are fixed in most detailed planning systems. So an overload on a work center can often be solved by temporarily modifying one of these parameters, such as reducing a batch quantity or using buffer stock. Now the demand on the resource will be less or moved to a different time period, one that can handle the demand.

In some extreme cases, it may be necessary to modify the sequence in the master schedule. This should be done only as a last resort, however, because changing the sequence in the master schedule will change the needed sequences on all work centers. This may cause more problems than it solves. Instead, try and solve the problems at the lowest possible level of product manufacture. Then execute the master schedule—don't change it.

Some newer, more complex detailed planning systems are now coming on the scene that eliminate peaks and valleys of work on work centers—at least for critical bottlenecks. OPT from Creative Output, Milford, Connecticut, OPIS from the Carnegie Group and IBM, and ISIS from Carnegie Mellon University and Westinghouse are examples of the new scheduling methodology being developed. They schedule flow rate and sequence at the same time and will modify the planning parameters automatically to ensure a level load on the bottlenecks and optimize output from them. All other work centers are scheduled to support the bottlenecks regardless of peaks and valleys.

This is logical, as the only work centers that need to be optimized are the bottlenecks. These control the total output of the factory.

The need for actual flow rate and sequence monitoring doesn't go away with this type of planning system. If anything, it becomes even more necessary. But at least these newer systems reduce the built-in peaks and valleys that result from simplistic planning systems.

FOLLOW THE HIERARCHY

The hierarchy of plans shown in Figure 4-1 is logically correct. It makes sense to deal with large levels of aggregation in the future and continually refine these gross numbers as time goes on. The key is to get the top numbers right. Then, you'll have a better shot at figuring the detailed numbers correctly.

The detailed numbers are important because you base your commitments on them. These numbers will prompt you to buy or make specific items, and commit company funds to particular projects. If these numbers are wrong, you haven't got a chance of meeting your company's goals.

The arrows flowing from the top plans to the bottom in Figure 4-1 are critical. Few companies have these as firmly in place as I would like. Evaluate your own planning process against this model to see if you are satisfied—or if there is room for considerable improvement.

The master schedule is a more powerful set of data than most managers realize. It is the interface between the generalities in the strategic business and production plans and the specifics in the detailed planning system. We'll discuss these critical numbers further in the next chapter.

5 | Planning Step 2: Create a Master Schedule That Gets the Right Product to the Right Place at the Right Time

All operations in a factory are designed to support the shipping of a product at some time in the future. The shipment plan may be based on a forecast, a booked customer order, or some combination of both. It's obvious, therefore, that your ability to do the right things today in operations is largely dependent on your ability to predict this future need to ship something.

We discussed the hierarchy of plans in the previous chapter. In this one, we will deal with long-range strategy and resource analysis up to the point where the master schedule takes over. The master schedule numbers are the key interface between strategy and the detailed operating system, and these numbers must reflect our best guesstimate about the future.

Master scheduling is a conceptual subject. Master schedules are unique to each company and product, and can vary based on the economy or today's business goals. My objective will be to define the primary concepts behind master scheduling, using clear examples. I'll summarize them all at the end of the chapter. Your job is to take these concepts and apply them to your specific business.

PREPARING THE MASTER SCHEDULE

My definition of a master schedule is, "A forward looking plan of what we intend to make, how much, and when." The key word is "make" because this has to be the plan to produce. It will rarely be the same as the plan to sell.

Now before all you sales and marketing people close this book in disgust, I had better explain why it's our plan to make things, not sell them.

Consider a company that makes toys. As you can imagine, toys have a very seasonal sales pattern. Seventy percent of annual toy sales occur during only two months of the year, September and October. This allows distributors and retailers time to plan their stocks to meet the huge Christmas demand.

As I am sure you can also imagine, the factory cannot produce 70 percent of its annual production in only two months. It has to produce on a more even basis throughout the year, at least in the capital intensive areas of the plant, such as plastic molding. It has some ability to flex final assembly using college students and temporary hires, but still not enough to match the seasonal sales.

It's obvious, therefore, that the sales plan in this case will differ from the master schedule. Since the operations people must have a level-loaded schedule for detailed planning, the master schedule must reflect this level production objective, not the highly seasonal sales. Inventories allow this decoupling of sales and manufacturing plans.

There are reasons other than seasonality for separating the sales and manufacturing plans. We'll discuss these other reasons as they arise.

Three Key Master Schedule Objectives

The master scheduler's job is to achieve three primary objectives, as detailed below.

1. *Convert aggregate production plans into the details of product and time required to drive the logistics activities of the business.*

   This conversion of production plan data into details was discussed in chapter 4 but is worth repeating here. A summary of the details in the master schedule should closely match the aggregate data in the production plan. Important differences must be highlighted, and executive managers must determine which is correct, the production plan or master schedule. All other detail plans affected by this decision must then be adjusted accordingly.

2. *Simultaneously support the marketplace and run the operations effectively.*

   I can make this seeming contradiction even worse if I add,

"be flexible to the marketplace and stable in the factory." But these are not conflicting ideas, as you will see later. Your skill at balancing these seeming contradictions will contribute a lot to your company's success.

3. *Ensure the demand stream (forecasts and customers' orders) fits within your production and vendor capabilities.*

This is the flow rate capability discussed in chapter 2. Never let the marketplace demand exceed your capability to produce. All kinds of bad things will happen if you do, chief among them critical inventory shortages and poor customer service.

Complete, Total Output Must Be Defined

The role of the master schedule is to define the total output of the plant as completely as possible. This latter phrase has two implications. First "as completely as possible" means all known production streams must be included. Finished goods, spare or service parts, affiliate plant needs, intermediate products, and special demands, such as for research and development prototypes, all need quantification in the master schedule. (Intermediate products are those items not processed to an end product. They could be subassemblies or semifinished products. This type of sale is very common in the process industry.)

Let's go back to the pipeline analogy we used earlier. The flow rate out of the plant can be depicted as in Figure 5-1. The proportion of each of these activities in the master schedule will depend on your specific company or product. But it is obvious that the actual flow rate through all upstream resources must be capable of supporting this total demand on the business.

Second, "as completely as possible" also implies that it is impossible to make the master schedule perfect. Before the ink dries on the paper, your numbers will be wrong to some degree, simply because you cannot be completely accurate when predicting the future. Forecasting inaccuracy must be considered when creating the master schedule and its influence countered where possible.

WHAT TO PUT IN YOUR MASTER SCHEDULE

The master schedule is an amalgam of several inputs, some unique to certain products, others of a more general nature. Some thoughts about the major inputs follow.

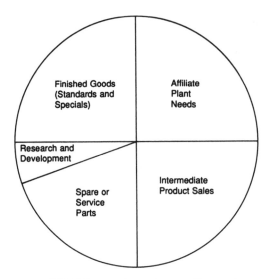

FIGURE 5-1. Total Factory Demand

Forecasts. Many manufacturers get most of their information about future needs from forecasts. They produce the so-called make-to-stock products, such as radios, televisions, watches, and hardware items. We'll come back to forecasting later in the chapter to understand the characteristics of forecasts and see how best to consider these when creating the master schedule.

Customer orders. A few manufacturers get all their information about future needs from already booked and promised customer orders. These companies produce true make-to-order or engineer-to-order items. Examples are ships, commercial aircraft, and government or military products.

Many manufacturers require a combination of forecasts and customer orders. Some of the future horizon is sold out to specific customer orders but the sold horizon does not extend far enough ahead for all detailed planning needs. Hence the horizon beyond the sold business is filled in with forecasts. Product examples include machine tools, telephone equipment, and many industrial products.

Planned inventory changes. It's obvious that if you plan to increase or decrease finished goods inventory levels then your master schedule must reflect this change. Let's say you've made a decision to increase finished goods, either because you have added warehouses or because you want to improve product availability. That means your master schedule must exceed your sales plan for a while to build the extra products. Similarly, if you plan to decrease finished goods, your master schedule rate of production must be less than your sales plan temporarily to bleed off the excess inventory. (Inventory change is another clear reason why the master schedule is based on what you intend to make, not sell.) Other inventory changes, not just finished goods, also impact the master schedule. We'll see their influence later.

Lead times. These words bring back into focus the P:D ratio discussion from chapter 3. If you remember, P is your total production and procurement lead time for a product: D is the customer's lead time for the same product. You must base certain specific characteristics of your master schedules on your P:D ratio relationship. In this section I am going to assume your P:D ratio is greater than 1 to 1.

Several lead times are important when creating a master schedule. At the moment let's focus on just one: the stacked, cumulative, or aggregate lead time, shown in Figure 3-1. This is the critical path lead time that begins when you place an order for raw materials and extends through the time they arrive in the plant, are processed through various stages, and manufactured into finished products ready to ship. We could also add engineering time ahead of the raw material procurement date for an engineer-to-order product and distribution time after the product is made for make-to-stock items.

As you can see, this length of time determines the minimum horizon for numbers in the master schedule. You can also see the logic of detailed planning from this picture. The objective is to put the right number in the master schedule at the minimum horizon position (15 weeks in the example) and use detailed bill of material calculations to figure out how much of the right raw materials to buy today.

I routinely ask in my courses, "How long is your stacked lead time?" I get answers ranging from 16 to 26 weeks to 50 weeks and more. It's a rare company or product that has a stacked lead time of less than 16 weeks. How long is yours?

Now that you know your lead time, when is that point in time from today? Please calculate it for your products and consider your ability to put the right number into the master schedule at that point in time. Also make sure you understand that your purchasing department is buying some raw materials and purchased parts today to support that number. If your prediction in the master schedule is good, then purchasing is buying the right things. If it's bad, they are buying things that will prove to be wrong. Excesses of some items and shortages of others will be the inevitable result.

Stop your people from quoting lead times. If they must quote lead times, force them also to quote the date the lead time implies. This will help them realize the value or futility of the data at this future point in time.

For make- or engineer-to-order products, the actual ship date of the product, even at the end of the stacked lead time, is fairly accurate. The problem in this business is that customer specifications or the design itself keep changing. So you wind up with the same problem described earlier: you are buying or making the wrong things. Excesses and shortages are inevitable in either case, not to mention the high costs of expediting the newly created shortages. Poor customer service, high inventories, and poor return on investment will be hard to avoid.

Lead time reduction is a key concept we'll tackle in chapters 8 and 9. If the stacked lead time for your products is reduced, you will have a lower P:D ratio, the predictions contained in the master schedule will be more accurate, and there will be less time or need for customers and designers to make changes. For now, though, we'll assume your stacked lead times are fixed and cannot be changed, and we'll figure out how to make a master schedule under these conditions. Then, reducing lead times will only make your master schedule better still.

Lot sizes. Many plants like to make large batches of products at one time for efficiency reasons. If the sales forecast is 10 units per week but manufacturing elects to make 100 every ten weeks, then it's obvious the master schedule must reflect 100 every ten weeks. Here is another clear example of the difference between the sales plan and the master schedule.

Capacity. Up to this point in our discussion, we have created what I call a "wish list" or desired master schedule. We have looked only at forecasts, booked orders, desired inventory changes, lead times, and lot sizes. The next question is, "Can we produce this amount with our current and future capability?" Here is where the total desired output flow rate of the factory, as defined in the master schedule, must be checked against the actual upstream flow rate capability. Chapter 2 discussed this at great length and we'll come back to the specific methodology for doing this in chapter 9. It's worth repeating: the master schedule must be a feasible plan, supported by adequate upstream capability. This test must not be shortchanged. Too much hinges on balanced flow rates.

FORECASTING: TWO KEY PROBLEMS AND HOW TO SOLVE THEM

Let's leave the master schedule discussion for a while and concentrate on the forecasting process. We'll return to the master schedule later.

Most companies—even make-to-order companies—need some forecast element in their master schedule to fill out the necessary planning horizon not covered by booked orders. Of course, make-to-stock companies base all their master schedule on forecasts.

The forecasting problem is really two problems, quite different in concept and approach. The first is, "How do you make better forecasts?" The second is, "How do you use forecasts more effectively?"

Stop Looking for a Magical Way to Make Better Forecasts

The first problem of forecasting has taken altogether too much of our time. We keep looking for the new magical statistical forecasting technique that will allow us to predict the future with great certainty. Unfortunately, such a technique doesn't exist. But we keep hoping. The track record is forecasts are either lucky or lousy. Even so, many people, when questioned about their poor business performance, answer, "If only we had an accurate forecast, things would be wonderful."

These two words, "accurate" and "forecast," don't belong in the same sentence. They are a contradiction, or, more correctly, an oxymoron, just like "military intelligence." Now don't get me wrong—I'm not saying that improving the quality of your prediction is a complete waste of time. But the law of diminishing returns applies very much to forecast quality. At some point, the forecast you've got is the best you're going to get.

Use Forecasts More Effectively

Since you can't get an accurate forecast, your question instead should be, "How do I use forecast data, with its embedded accuracy problems, most effectively to run my business?" I don't think we have taken as aggressive an approach to this line of reasoning as we should. It holds enormous promise for making real improvements.

The Facts and Myths Behind Forecasts

Both ideas, making better forecasts and using them more effectively, are an integral part of creating a master schedule. Few companies are doing this well. I'll prove that to you with some true statements about the characteristics of forecasts.

Forecasts are almost always wrong. This is a true statement if you are looking for perfection in the forecast numbers. If you expect error and believe the forecast is only an approximation, then this statement is probably false.

Forecasts should include three numbers. The three numbers are best guess, upper limit, lower limit. If we expect forecasts to be wrong to some degree, then the data from them should give us only a sense of direction plus some estimate of the expected error. The error range could come from past experience of forecast error or from marketing inputs.

This statement is obviously true. You cannot forecast with precision. Now the question is, how do you use three wrong numbers in the master schedule to get the right result for the business? We'll discuss that idea in chapter 6.

Forecasts are more accurate for families. Any time you aggregate data, the laws of statistics say the data quality improves. The forecast for the total annual sales of your company in dollars is a reasonably accurate prediction. The forecast of sales for a specific product is much less accurate.

But what is a family? It's obviously a grouping of products, but based on whose criteria—sales and marketing or manufacturing? A sales and marketing family could be based on customers or market. A manufacturing family probably will be based on common resources or raw materials used to make this group of products.

Things would be perfect if both groups agreed on common families. If they cannot, however, you must come up with a "translation process" between them.

Forecasts are less accurate further out. This is another law of statistics: the further into the future you try to predict a specific event, the more wrong you are likely to be. This statement is true, and it reaffirms my earlier statement about lead time reduction. The shorter the stacked lead time the better.

Ignore These Myths and Facts at Your Peril

Although these myths and facts of forecasts are well known, few master schedules pay much attention to them. One of my clients, not atypical, used to predict specific end product sales six months into the future, six months being his stacked lead time. They forecasted only a single number, without ranges. Then, they used this one number, projected six months ahead, to create their master schedule for specific finished goods. You can be sure they had excess inventories and shortages at the same time.

Sillier still, their finance department required forecasts of sales for specific finished goods in September of one year for sales in the following fiscal (also calendar) year. Finance used these numbers to create budgets. Both practices are crazy. You can't predict this level of detail that far ahead of time with a high degree of accuracy.

Remember, closing your eyes to these forecast characteristics won't make them go away. You have to learn how to build them into the planning process. Forecast characteristics have broad implications for designers, the manufacturing processes, the detailed planning data, and financial budgeting processes.

THE CRITICAL ZONES OF PLANNING

Just as there is a hierarchy of plans for the total business as described in chapter 4, so is there a hierarchy in the master schedule. The future cannot—I probably should say must not—be one homogeneous set of data. The forecast characteristics discussed earlier should convince you of that. But since the advent of computerized operations systems, there has been a tendency to think about the master schedule as one set of data. We need to think instead about data to suit a given time period.

This concept is illustrated best by the picture of a moving scroll shown in Figure 5-2. Time winds onto the left spindle and becomes history. Time unwinds from the right spindle to give us a constant view of the future.

Over the short range, we must be both product and time specific in our planning because here we are producing specific items to go to customers or warehouses on specific dates. And forecast accuracy for the short range can probably provide fairly correct and adequate information.

But intermediate-range plans can't include the same level of definition in terms of either time or product. Forecasting is not accurate enough to provide good information this far out. But these numbers are only being used for purposes of scheduling components and subassemblies, not finished products. (If this point is not clear, refer to Figure 3-1 to see how these numbers are used.) If these components and subassemblies are common to wide ranges of end products, then

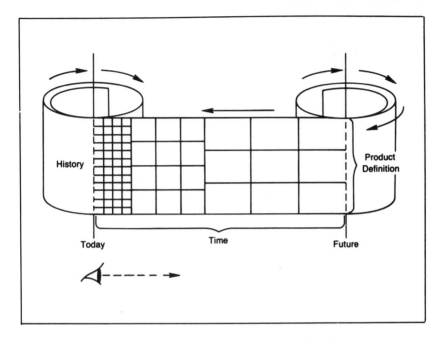

FIGURE 5-2. The Future Planning Scroll

a master schedule for families of products is specific enough, drawn up by week or even month. Forecast accuracy at this level of aggregation will compensate for the lengthening time horizon. But now the design implications for the product become a significant factor, as we will see in chapter 6.

Over the long range, the numbers are used only for raw material and long lead-time item procurement. If these items and materials are standard to large product groups, even larger aggregations of product can be incorporated into the master schedule, with even longer time increments.

A company in Northern Italy that makes white goods has really grasped this concept well with their QME plan. "Q" means quarterly planning, done long range in gross product groups, for example, freezers, dryers, washing machines, and similar appliances. These numbers provide the basis for buying raw materials, which are controlled so as to be common to each product group.

"M" means monthly planning, done only for a few months out. This planning includes more detail and applies to the product family level, for example, gas versus electric dryers. The company uses this level of detail to buy specific components and schedule production of subassemblies from the common raw materials.

"E" is execution. Projecting only a few weeks into the future, the company defines the specific end products they intend to assemble from the subassemblies and specific components built during the "M" period. The execution plan is tailored to suit actual warehouse stocks and known customer orders.

You can see the impact on product design such a planning methodology demands. Long lead-time raw materials must be common to large product groups. Subassemblies and specific components must be common to product families. End items must be quickly producible from these standard subassemblies and components. A logistically friendly design is a prerequisite to their success.

FIVE KEY CONCEPTS OF MASTER SCHEDULING

Master scheduling is a conceptual subject, not a scientific one. Let's review the five key concepts behind master scheduling to make sure they are clear. Your next step will be to select those concepts that apply to you and develop a specific strategy for incorporating them into your own master schedule.

1. *The master schedule is not the sales or shipping plan.* This is obvious when you consider seasonality, planned inventory changes, and lot sizing.

2. *Only define in the master schedule the degree of product detail that is forecastable.* All master schedules are future plans, hence most of them will have to rely on forecasts somewhere in the planning horizon.

Finding the best level of detail to forecast is critical, as you can see from our discussion of the critical zones of planning. Product design and bill of material structures must all be developed with this in mind.

3. *You must take into account the facts and myths about forecasting.* We discussed four forecast myths and facts earlier in the chapter. They are fundamentally valid, so you cannot ignore them. Build your master schedule with them in mind.

4. *The master schedule must contain contingencies.* This is especially the case if the P:D ratio is greater than 1 to 1. The speculation necessary to provide an adequate planning horizon demands some

contingencies, either safety stock or overplanning, to counter the inherent forecast error.

5. *The master schedule must provide stable factory and vendor schedules but also be flexible to the marketplace.* This seeming contradiction can be handled with a clear understanding of the difference between the master schedule and the final assembly schedule (discussed in chapter 6), product designs that assist planning, and judicious contingency provisions.

6 | Planning Step 3: Design Products to Give Customers What They Want—When They Want It

The P:D ratio discussion from chapter 3 and the master scheduling discussion from chapter 5 both alluded to product design as a help or hindrance to effective logistics. The P:D ratio dilemma of Figure 3-5 shows the problem clearly. You must forecast that portion of the P time that is longer than the customer's D time. If your forecasts are fairly accurate, high customer service, low inventories, and a low-cost business will be the happy result. But if you can't forecast this horizon accurately, then the business results will be devastating.

We have already alluded briefly to the three actions you can take to combat this problem. They are

1. Simplify the product line.
2. Standardize the product.
3. Overplan to add contingency to cover the forecast error.

We will look at these ideas in more detail in this chapter and make their link to product design crystal clear.

The result should be a decision on your part to develop a design architecture within which product designers are allowed to work.

Your objective: logistically friendly designs that will significantly improve your business results.

WHAT'S THE PROBLEM WITH MOST PRODUCT DESIGNS?

We can use Figure 6-1 to further explain the planning dilemma and suggest some solutions. It shows a hoist, made up of one hook, two control pendants, four gear boxes, ten drums, and thirty motors. The number of end configurations is 2,400 ($30 \times 10 \times 4 \times 2 \times 1$). The actual production rate in this example, based on the aggregate sales forecast, is 50 per week and the stacked lead time is 15 weeks, just like Figure 3-1.

The obvious question is, which 50 out of 2,400 possible hoists will customers buy 15 weeks into the future? If we can figure that out, then we can easily make a master schedule and proceed with our detailed planning. The problem, of course, is that faced with 2400 choices, you cannot predict with any certainty which 50 hoists customers will buy in each of the next 15 weeks.

This problem can be seen in its extreme in automobiles. The major American automobile companies can theoretically produce far more varieties of cars than their actual production volume of cars in any year. How do you schedule in a situation like this?

Part of the answer lies in understanding the specific P:D ratio for a product. Returning to the hoist, the P time is 15 weeks as stated earlier. The D time is 4 weeks and is competitive. This means that procurement and production over only the first 11 weeks of the production process have to be based on speculation.

If you look at the hoist figure more closely you see it is made up

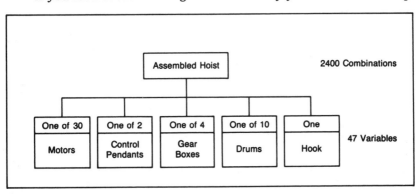

FIGURE 6-1. Product Definition to Improve Forecasting

of five different modules. Maybe we could predict demand for these five modules fairly accurately and have them flowing toward the final assembly area. When the actual customer order comes in, all we have to do is mix and match the modules in final assembly, test the hoist, and ship. If final assembly and test can be done in four weeks, then there is no need to try to predict which 50 hoists customers will buy 15 weeks into the future. We only have to predict how many of each motor, drum, gear box, pendant and hook will be ordered. Only 47 predictions are required (30 + 10 + 4 + 2 + 1) 11 weeks into the future, much easier than 2400.

Predicting the future at the end product configuration is impossible. The combinations of options drive the possible number of finished goods much too high. But at the module level the multiplication effect of the options doesn't apply. They can simply be calculated arithmatically, so hence predicting the future becomes a much easier job.

Look at the power of this idea. If engineering designed an alternative hook (God bless them), the possible hoist combinations would double to 4,800. But only one more prediction would be required at the module level.

As you can see, standardization is critical. This concept works because of standardization in the product design. Without it, you'd be faced with an impossible forecasting job and hence, poor financial results. But many designs are not standardized. They have unique parts which require long lead times, a sure road to disaster. We'll elaborate on this point using some case studies later in the chapter.

Master Scheduling the Hoist

Let's explore the hoist example in more detail. It's obvious that with only one choice, the hook's master schedule is going to be 50 per week. This is based on the aggregate sales volume of all hoists regardless of specific configurations.

The pendant comes in two varieties, A and B. We know the total should be 50 per week, the aggregate sales rate. Past history or a forecast of mix could split this 50 for us into the share applied to each pendant. Figure 6-2 shows we have used a 60/40 split in our example. This puts 30 pendants A and 20 pendants B into the master schedule. The same method could be carried over to the gear box, drum, and possibly the motor. However, as the varieties of each module grow, this method becomes less workable. For example, you have

Product	Average Percentage Mix	Master Schedule Quantity
Pendant A	60	30
Pendant B	40	20
Totals	100	50

FIGURE 6-2. The Pendant Mix Split

30 choices for the motor but since you are making only 50 hoists per week, the percentage splits are going to be very small.

Five Ways to Handle Excess Variety

The problem of predicting which 50 hoists out of 2,400 possible end products has been improved by focusing on modules. But what can be done when an individual module such as the motor has excessive variety? There are five choices.

1. *Offer fewer varieties.* Reduce the offerings to a predictable number. Maybe this could be done by selecting universal motors that handle a range of power inputs. Your marketplace offerings are the same, but the variety of products you need to support the marketplace is less, improving your chance of having the right product in stock.

This may increase the standard cost of the end product, but it could be worth it if it improves customer service at the same time it reduces inventories. This trade-off of product cost against flexibility and lower inventory will be discussed in more detail later.

2. *Break the product down further.* Evaluate the motors and see if the combining of options problem we explored with the hoist also applies here. Maybe we'll find five voltage variations with three mounting methods with two stopping choices, without brake or with brake. Multiplied together we get 30 end products. But if the design is such that the voltage sensitive parts are all identified and grouped into five bills of material, and the same is true for the mounting and brake sensitive parts, then we need to make only 10 predictions for the motor (5 + 3 + 2) not 30. For this to work, we have to be able to assemble the motors, connect the motors to the hoists and test the finished product in four weeks.

Look at what this has done to the prediction problem. Instead of 47 modules to forecast we have only 27 (10 + 10 + 4 + 2 + 1). We are moving in the right direction.

How can you make sure your designers think about common modules as they design products? By providing a design architecture directing them to design for logistics.

3. *Reduce the item's P time below 4 weeks.* It is obvious that if the total P time to produce motors and assemble them into hoists is less than 4 weeks, then the prediction problem disappears. We make these items to order, no speculation required.

4. *Offer differential lead times.* You could offer hoists containing a limited selection of motors the short lead time of 4 weeks, the hoists needing unusual motors get a longer (15-week) lead time. In the latter case you are pushing the D time out to meet the P time because of the speculation problem. This is a fairly common occurrence in industry. But be careful you aren't pushing your customers so far into their speculation horizon that you are beset with specification changes.

5. *Hold risk inventories.* Because of the unpredictability of the unusual motors, hold a minimum stock of these. Replenish based on usage. This will give you some slow-moving inventories but may be worth it for the improved customer response you can provide.

THE CHOICES ALL RELATE TO LOGISTICS

As you can see, these are logistics decisions and impact all business functions. Standardizing the motors and making the hoist a modular product involves design engineering. Getting short P times also may involve design engineering and certainly includes industrial engineering, purchasing, and manufacturing. Sales and general management may also have to get involved at least to some extent in order to cut P time.

Segregating often ordered from infrequently ordered products and offering different lead times for each certainly involves sales. It could also involve finance and general management.

Carrying risk inventory on the unusual motors to give adequate market response also has sales and financial implications. And, of course, assembling hoists in four weeks has implications for the factory.

The key thing to realize is you need a logistically friendly design that will allow you to achieve your business objectives. This is a crucial area that has not received enough attention from designers in the past. Far too many products are logistically unfriendly, even though they perform well, and are at the root of many companies' operating problems.

HOW TO CONFIGURE THE END PRODUCT

Figure 6-2 showed the separation of end products into modules to help ease the prediction problem. This means that the bills of materials that define the hoists must also be segregated so all components unique to a given module are grouped into one bill of material that defines that module. This is called "restructuring" the bill of material.

All we have now is a number of small bills of materials, 47 in our example, that define only the modules. How do you create a bill of material that defines a specific hoist?

The best way is through a menu selection sheet, shown for the hoist in Figure 6-3. It's called "menu selection" after a restaurant menu, a good analogy to use to understand the problem clearly. In a restaurant, every customer could conceivably order a different meal. The combinations of the various courses, foods, methods of cooking,

ITEM A—CONTROL PENDANT		ITEM C—DRUM	
Without Emerg. Stop	(1)	25 FT Capacity	(1)
With Emerg. Stop	(2)	50 FT Capacity	(2)
		75 FT Capacity	(3)
		100 FT Capacity	(4)
		etc.	
			(10)
ITEM B—GEAR BOX		ITEM D—MOTOR	
10 FT/MIN Line Speed	(1)	Without Brake—240V 60 Hz	(1)
20 FT/MIN Line Speed	(2)	Without Brake—440V 60 Hz	(2)
50 FT/MIN Line Speed	(3)	Without Brake—550V 60 Hz	(3)
100 FT/MIN Line Speed	(4)	With Brake—240V 60 Hz	(4)
		etc.	

650 A1 B3 C4 D2
(30)

FIGURE 6-3. Hoist Menu Model 650

and beverages are enormous. But by creating modular structures a restaurant can predict food usage long term, create the finished product specifications with the help of the customer (via an order slip), and produce the finished product in a matter of minutes.

The menu sheet for a manufacturer could be used as a sales order entry sheet, so customers help build the specifications of their specific products. It could be a sheet or video screen used internally by accounting or engineering to get a picture of the end product and its associated makeup. It could also be used by planning and scheduling to issue work orders to the factory to produce specific hoists, either to suit customer needs or stocked as finished goods.

The menu sheet from Figure 6-3 lists some of the 47 variables that can be selected to produce a finished hoist. The bottom line of the sheet shows a typical selection. Module A, the control pendant, comes in only two varieties: without emergency stop or with emergency stop. The example shows a selection without emergency stop. This means that later selections for the motor, module D, are now limited to only those without brakes. Further selections have been made for the gear box, drum, and motor.

The bottom line of the sheet is now analogous to a waiter or waitress's order slip. It defines exactly what configuration of product you need out of the possible options. Entering this into a computer bill of material configuration program will create the exact bill of material for this specific hoist.

HOW TO DRAW UP THE FINAL ASSEMBLY SCHEDULE

The final assembly schedule (FAS) or finishing schedule is another concept important to master scheduling. I have mentioned this term a number of times.

Think of the master schedule as a vacuum cleaner, bringing raw materials and purchased items into the factory and processing them up to the point where the D time starts, as shown on the left side of Figure 6-4. The final assembly or finishing schedule now takes over, shown on the right side of Figure 6-4. The FAS is based on either booked orders or warehouse replenishment orders. This schedule converts the items and materials provided by the master schedule into finished goods. Another good analogy is a relay race. Here, the baton is passed from the master schedule to the finishing schedule where the D time starts. The objective is to make the finishing schedule very

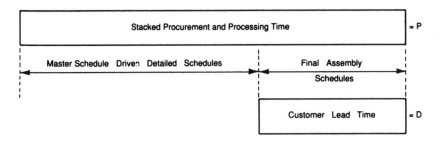

FIGURE 6-4. Master Schedule Versus Final Assembly Schedule

responsive to market conditions while at the same time keeping the master schedule stable.

Plan Modules, Produce End Products

The hoist example shows clearly the difference between a master schedule and a final assembly or finishing schedule. The master schedule for the hoist is in modules. Pendants, motors, drums, gear boxes, and hooks are all planned to a base rate (50 per week), and the mix is split based on historical ratios.

The final assembly schedule is driven by customer orders. These specify the particular mix of options for each given order. The master schedule provides the options based on historical ratios, the final assembly schedule consumes the options based on actual orders.

Another good example of this methodology is a bearing company. Bearings can be grouped into families in which the raw materials and components are largely standard. These families are called "basics" and have bills of materials that list the common raw materials and standard components.

Specific configurations of bearings can differ based on fit (the tolerance between the outer and inner race and the rolling elements), grease, packaging, seals, dirt shields, and so forth.

The master schedule covers production of the basic bearing and drives material procurement, machining, heat treating, and grinding. At some point—a well-defined time fence—this master schedule for a quantity of basics must be broken up into specific quantities for each configuration of finished bearing. This is where the final assembly schedule takes over and defines the exact configuration of bearing to assemble. This determination is based on customer orders or short-range, warehouse replenishment decisions.

FOUR SOLUTIONS TO POOR PLANNING PROBLEMS

I am sure several of you are wondering what happens if the actual customer orders don't match the historical split of options. Four things can happen.

1. *Customers don't get what they want.* The mix of options available to satisfy customers doesn't match up with the orders booked. In this case, some customers must wait and you carry excess inventories of unwanted options.

2. *A match is forced.* Give-away programs, discounting, and promotions persuade customers to buy the products you produced. This is very common in the automobile market, where incentive programs are offered for some car models.

3. *The plant scrambles.* As the actual order book differs from the plan, the plan is dynamically changed short range. But as you are now changing the plan within the P time, vendors and the factory have to produce the customer selected options in less than normal lead time. Expediting, air freight, and overtime can sometimes close the gap between what you planned and what your customers bought.

4. *Overplanning cushions the error.* I haven't described overplanning well yet but will later in this chapter. In simple terms it is safety stock deliberately created to handle the difference between master schedules and actual orders. If it is successful, then excess option inventories will allow customers to buy a mix different from that forecasted and get quick, reliable deliveries while the factory's schedule stays very stable.

MUSHROOM-SHAPED DESIGNS:
A KEY ELEMENT BEHIND LOGISTICALLY FRIENDLY PRODUCTS

The P:D ratio cries out for a mushroom-shaped design, as in Figure 3-7. The hoist has such a design as do bearings and automobiles. Mushroom-shaped designs go a long way to solving the automobile problem of theoretically more variety than is actually produced.

The white goods factory in Northern Italy discussed in chapter 5 has created a mushroom design for the appliances they produce. It has specified materials common to wide product ranges and subassemblies common to product families. The actual definition of end products occurs as late in the production process as possible.

This mushroom concept is a critical one for all designers to grasp. It is one of the key elements of a logistically friendly design. We'll see later some logistically unfriendly designs and the unfavorable results they caused. Discussing how to make them logistically friendly will reinforce the key concepts. You should now be ready to define your own control mechanism on future designs, that is, your unique design architecture. This will guide your design engineers into generating logistically friendly designs in the future as a matter of routine.

WANTED: A LOGISTICALLY EFFECTIVE DESIGN ARCHITECTURE

Some of the concepts already discussed, such as the P:D ratio, forecast characteristics, final assembly schedule separate from master schedule, and mushroom buildup, all point to a specific design philosophy. But few designs fit the correct mold. The major problem is the set of objectives most designers have to meet.

Six Key Objectives for Designers

Most design engineers will tell you they have three objectives to meet with any new or revised design.

One, it must perform a desired function;

Two, at a target cost; and

Three, it must be aesthetically pleasing.

Rarely, however, do the objectives also include:

Four, serve the customer with maximum product availability;

Five, with minimum inventories; and

Six, with maximum flexibility to changing markets.

But the inherent design of a product controls 80 percent of your ability to perform to the last three objectives. And for most companies, these last three are at least as important as the first three.

FOUR CASE STUDIES OF DESIGNS GONE WRONG—AND HOW THEY CAN BE MADE RIGHT

Designs based on the first three objectives alone can be quite different from those based on all six. I'll show this in each case study. First

I'll present a design created using objectives one through three, then show how the design changes as I add objectives four through six. The questions I hope you keep asking yourself are, "How are my company's designs created?" and "What steps must I take to ensure all six objectives get equal consideration?"

1. Designing Logistically Friendly Pendants

The hoist example, Figure 6-1, shows two pendant choices. The specific functional difference is that one pendant contains only two buttons, one each for up or down movement. The second pendant contains three buttons, one each for up and down movement, the third to activate the brake on the motor that will stop the hook exactly at the right place. The second pendant is designed for hazardous use or for delicate positioning of the load.

If the designer considers the first three objectives only, chances are he will create two control boxes, one large enough for two buttons, the other larger to accommodate three. This design will require two different cover plates and probably two control cables. Commonality between the two pendants will be limited to two switches. Everything else will be unique. High inventories, poor product availability, and inflexibility to market shifts have been preordained as a result.

If on the other hand we can overlay this design with the last three objectives, the designer will create only one control box large enough for three buttons. Now, only one cover plate is necessary, with a plug button for when the third switch is not needed. The control cables are identical. The unused wires are taped out of the way when the brake is not selected.

Low inventories, high product availability, and flexibility to market shifts are easily accommodated with such a design. It is a mushroom, because all variability is added at the last moment. It performs the needed function and can be as aesthetically pleasing as the original design. The only concern I am sure you have is the cost element, objective two. Has this objective still been met?

I'll discuss the cost problem later in the chapter and address it again in chapter 10. In thinking about the subject, though, please consider hard versus soft costs. Hard costs are those that usually show up as product standard costs, rarely more than 50 percent of the total costs of the business. Soft costs are those we traditionally put into burden, overhead, staff support activities or items we don't even try to cost, such as inventories, poor customer service, and inflexibility.

You must consider the total costs when making design choices, not just the 50 percent in product standard costs.

Modifying the pendant shows how standardization of a design can really be put to use to help the business. The mushroom concept comes through clearly.

2. Designing a Hoist Shaft That Makes Forecasting Possible

The shaft connecting the hoist motor to the gear box is interesting. If the end that connects to the motor is unique to each motor, if the length is variable based on the drum selected, and if the end connected to the gear box is unique to each gear box, then there are 1,200 (30 × 10 × 4) possible shaft configurations. This makes for an impossible forecasting job. Of course, standardization, at least of the end connections, will simplify the problem.

But what if this cannot be done? In that case, making the shaft out of three pieces could be the solution as shown in Figure 6-5. Forty-four pieces (30 + 10 + 4) connected together in a variety of ways will produce the 1,200 shaft configurations and minimize the forecasting problem. The actual assembly of the shaft will occur only after the customer's order is received, i.e., during the D time.

A three-piece shaft would be a very unusual design for a designer to choose. It adds complexity and standard cost. But the payoff is that it certainly helps him meet the last three objectives we have set for product designs. The revised shaft design also fits into the concept of creating master schedules separate from final assembly schedules. With this design you can plan the 44 pieces long term and assemble the right 50 shafts out of 1,200 possible variations at the last moment.

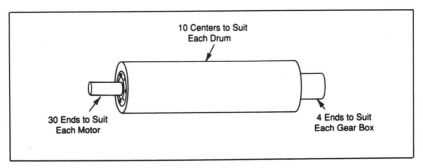

FIGURE 6-5. Alternate Shaft Design

3. Consumer Electronics: The Top Performing Product That Nearly Sank a Company

A European company designed a new line of stereo equipment. Its performance was far superior to the competition, costs were about the same as the competition, and aesthetically it was excellent. The stereo magazines rated it the best in the world.

That's the good news. Now for the bad. The product design was logistically unfriendly as diagramed in Figure 6-6.

The product range consisted of 56 end configurations. Some of these were color and foreign language variations. Twenty-two different subassemblies, specifically printed circuit boards, could be assembled into these 56 varieties. Six bare boards could be assembled into the 22 subassemblies through the selection of some unique components.

The lead time (P time) in months to procure, produce, and distribute this product are shown on the left of the diagram. Point zero is when the end consumer buys the product. The distribution time from the factory to the consumer averaged 3 months, during most of

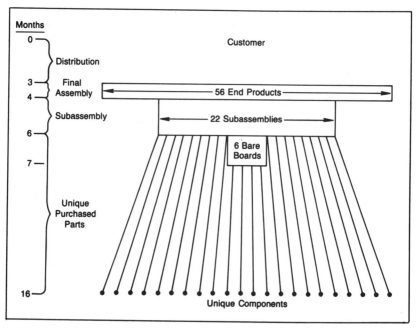

FIGURE 6-6. Time-phased Product Makeup

which inventory simply sat on the shelves of various warehouses and distributors. Final assembly took one month, subassembly took two months, and obtaining the unique purchased components, available only from one source in the Far East, took 10 months. The bare boards took one month. Hence the total "P" time was 16 months.

The first process step in the subassembly production was to add the unique components. Hence the six common bare boards were made unique 6 months before the customer purchase. The 22 subassemblies were made unique to the 56 end products in the first process step of final assembly, so that color and language had to be predicted 4 months before customer purchase.

I am sure you can envision the logistics problem. How do you predict the right unique components 16 months before the customer buys? In fact, the company couldn't. How do you predict which 22 subassemblies to produce six months ahead of demand? Again, the company couldn't. And worst of all, it couldn't predict which of the 56 end products to make 4 months ahead of the sale.

The end result of this design was huge inventories of some end products and backorders for others. And with stereos, as is the case with most consumer products, backorders meant lost sales as consumers switched to the second or third best competitive product available.

In my opinion this company should have billed the competitors for marketing expense. They stimulated demand with a product that performed exceeding well, but they couldn't fulfill the demand for it. The competitors reaped the benefits of an outstanding but logistically unfriendly product.

Not only were inventories of finished products high and unbalanced, but so were the inventories of subassemblies and unique components. The number of predictions needed plus the length of the forecast horizon guaranteed it.

I am sure you can visualize the frantic activities in the company as they tried to get out of this condition. Expediting, air freight, overtime, building products with missing parts and then later adding them, idle time, and so forth were all rampant. These "soft" costs were all incurred in addition to the predicted hard costs for the product.

A discussion with the designers showed that the product could have been designed to meet the performance, cost, and aesthetics objectives as well as being logistically friendly. The revised design is diagramed as Figure 6-7.

The 56 end products now take on their unique identity at the

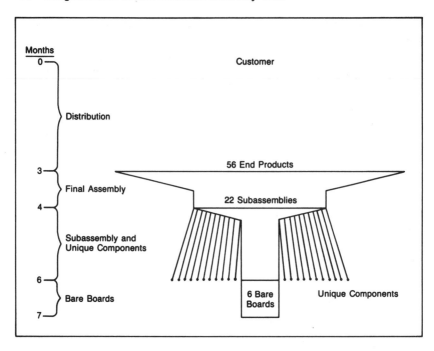

FIGURE 6-7. A Logistically Friendly Design

last step of the assembly process. This cuts one month out of the prediction horizon. The 22 subassemblies are also made unique at the last process step, cutting two months out of their prediction horizon.

The unique components are now selected from standard products, cutting their procurement time to 2 months. The total P time between ordering these components and selling the end product to the consumer is now 6 months compared to 16. If the company had done all this in the first place, just imagine what kind of success they'd have had in the marketplace.

This is an example of how the P:D ratio must be considered in a product design. P times are especially important for products with short life cycles. In this case, consumer electronics, life cycles are rarely as long as two years. Having a "P" time of 16 months with a life cycle of two years is fundamentally an invalid way of operating a business.

4. Industrial Products: Standardizing Designs for Specialized Markets

A company in America makes scanning technology used to monitor pulp and paper mills, aluminum foil production, cigarette machines, galvanizing lines, and so forth. It builds a variety of scanners to measure the critical characteristics involved in these processes. It collects these readings and either displays them for human analysis and action or analyzes them automatically and directly controls the production process.

The problem is that none of the application markets is very large. Ordinarily the development costs incurred in designing products for each market would have to be spread over limited production runs. This means either the costs per installation would be prohibitive or profits would be difficult to come by.

To solve this problem, the design engineers analyzed the product and sorted out what had to be unique to each market and what could be standardized. Their answer was that the scanners had to be unique but the data collection and analysis electronics could be standardized.

The standardized electronics would be more complex than any one application needed. With only one electronics unit for all applications, however, the total development costs would be lower. The designers also found that other support costs would be lower, for example, planning, scheduling, purchasing, and accounting. They used this strategy on the electronics. They are now trying to develop the same concept for the mechanical parts of the products that support the scanners.

Compare this to a medical equipment company that makes X-ray diagnostic equipment for hospitals. The company has separate design teams for each product type. Even the tables on which the patients lie are different for each product, as is the display unit the doctor sees. Many of these, and other items, could be standardized among product types. But this would mean reorganizing the design efforts along modules of products, for example, tables and display units, instead of teams per end product.

Because of the company's nonstandard approach profits are hard to come by, spares support for such a vital product is very costly, flexibility to market needs is poor, and inventories are high.

Here are two good examples of constrained versus unconstrained design activities. Don't forget that design engineering consumes only 5 percent of the total life costs of a product—but this 5 percent commits you to 80 percent of the total life costs, Don't let the 5 percent you spend be uncontrolled. Constrain it to give you the total results you are after.

DON'T BE PENNY WISE AND POUND FOOLISH BY OVERLOOKING SOFT COSTS

A large part of the problem with product designs lies in one of the first three objectives of designers, product costs. With this as a constraint, they try to design to minimum cost.

But true product costs, as mentioned earlier, cover up to only 50 percent of the total costs of a business. Other costs are allocated to products on some arbitrary basis, for example, factory burden. Some are gross margin deductions, such as sales expense, and never end up as product costs. Still others, such as lost market share, don't show up anywhere as product costs.

The true product costs of direct materials and unburdened direct labor are what I call "hard" costs. The other costs are the "soft" costs.

It is obvious that a design must minimize the total costs to the business. But total cost information is rarely available to designers. And because the soft portion of the costs lacks the precision and solidity of the hard costs, design decisions are usually made based on minimizing the hard costs only.

To illustrate the problem of focusing only on hard costs, let's look at a valve manufacturer in the Midwest, which has a direct labor burden rate of 550 percent. The design engineers used to specify more expensive materials to avoid direct labor which carries the punishing burden rate. Hence, the product standard cost was reduced! But the burden didn't go away because most of it was fixed cost. So, because direct labor hours were reduced, the burden rate increased. Because they are now buying more expensive materials, total costs actually increased even though by financial calculations they appeared to be reducing. They recently revised their complete design activities around total costs, with huge benefits to the business.

A Second Look at a Costly Mistake

The consumer electronics company we discussed earlier also had troubles with hard and soft costs. After the engineers described how they could have redesigned their stereo system to make it logistically friendly, I asked them how much additional cost they would incur. Their estimate: 85 cents per unit, not a negligible amount for a high volume product. This was equal to about 1 percent of the total product cost.

Now the question. Should the company have spent this additional amount? In retrospect, the answer is yes. The poor results this company achieved would have been completely reversed if they had spent an additional 85 cents per unit. The actual additional costs per unit turned out to be several times this much because of the expediting, air freight, idle time, and workaround activities they had to perform. On top of this the company lost sales and had customer complaints and a demoralized work force to contend with. It was the most expensive 85 cents this company ever saved.

How to Prevent Mistakes Like These
In Your Own Business

The above example clearly illustrates the conflict between hard and soft costs. But the solution became apparent only after the damage was done. How do you prevent poor product designs before the fact? The answer: By establishing a design architecture designed to achieve product characteristics that are desirable from a logistics viewpoint. If you try to justify each design alternative on soft versus hard costs, you'll spend more time arguing about the accuracy of the soft costs than managing the business. Reserve the cost justification process for significant hard cost increases to provide logistically friendly products. Analysis by exception is the correct process to use.

A SIX-STEP STATEMENT OF DESIGN ARCHITECTURE REQUIRED FOR EVERY MANUFACTURER

Product design plays a crucial role in the success of a business. As stated earlier, only 5 percent of total life cycle costs are spent in developing a new product, but they commit a company to 80 percent of the total life cycle costs for the product. Hence, product designs must be constrained to minimize total life cycle costs. The emphasis should be on total costs, both hard and soft. Few designers are constrained in this way. Analyze your most recent products in light of this chapter's discussion to see if this is the case for your company.

Your statement of design architecture must address all the logistical problems we've covered. It is a guide to help design engineers pick the right design for the success of the product whenever they are confronted with design alternatives. It also precludes long discussions about hard versus soft costs. Only significant hard cost increases should have to be justified against their soft benefits.

Step 1: Develop an Accurate Market Needs Statement

Any product can have a wide or narrow market. The breadth of the market and how much of it a product is designed to reach must be clearly defined as a first step.

For example, electronic products can be designed for: consumers and/or professional people; high end, middle, or low price ranges; for multiple countries or only a selected few; and so forth.

Industrial products can be designed for all possible applications or limited to certain specific industries. The scanner company mentioned earlier clearly defined those industries it considered its market areas. This focus on key market areas allowed the designers to create a single electronics package to satisfy these markets.

This is a difficult but necessary first step. Looseness in your statement of market needs will result in an overly complex design with many add-on features as the market asks for them. These add-on features will be difficult, if not impossible, to keep logistically friendly. Product proliferation will doom such a design.

For example, the hoist pendant (described earlier) could have started out with only one choice, the up and down buttons. The small box and cable with few wires would have been sufficient. If later the need for the brake became apparent, then the small box would have been inadequate. The larger box and cable with additional wires would have been needed. Two pendants would have been designed, tooled up, put into the product catalog, and supported with inventories when only one could have done the job. These additional costs for two products instead of one must be recouped in the product price or paid for with reduced profits. The market needs statement is a critical first step to avoiding these costs.

Step 2: Consider All Six Product Objectives

The six objectives for product designs were defined earlier. You must insist these be clearly understood by the design engineers and everyone else influencing design decisions. Make sure these six objectives receive equal weight and beware the hard versus soft costs conflict.

Step 3: Develop Multiple Use Products

A broad product line is a difficult one to manage. One method to combat this is to design products that have application variety without product variety. An easy example is a medical electronics company that sells its products in America and Europe. Originally the company duplicated the product line for voltage differences on each side of the Atlantic. It never had the right ones in stock.

The company's new designs are dual voltage. A simple switch converts them from one power source to the other. The product line has been reduced by 50 percent, but the application variety has not been affected. The logistical friendliness of this design has been significantly improved.

The scanner company mentioned earlier took the same approach with its electronics data collection and analysis unit. A slightly more complex unit than any one application needs allows the company to serve a wide variety of applications with one unit. Forecast accuracy, flexibility to changes in demand between markets, and inventories to support sales are all remarkably improved.

Step 4: Control Component Selection

Components are often needed in an early part of the P time for a product. Their selection is therefore critical.

Control them within the following limits:

Standard— readily available from a wide variety of sources.

Common—Used across a wide variety of products.

Quickly available uniques—All items unique to a product must have a very short lead time.

Step 5: Design for Short P Times

The length of the lead time of procurement and production must be a major design concern and will often have to be considered along with the manufacturing process and the procurement alternatives. Short P times inherent in the design of the product can only pay dividends to the company.

Step 6: Create Variability at the Last Possible Moment: Grow Mushrooms

Designs must be standard throughout the early part of the procurement and manufacturing process. All the variability should be added as late as possible. A good example of this is dishwashers. Some companies include in the shipping box plastic panels of different colors. The consumer selects the color she wants, inserts the panel at the front of the machine, and discards the rest.

Compare this to earlier designs when color was a factory controlled variable. The finished goods inventories were always out of balance, simply because the right colors were never in stock.

Adding the variety at the end of the process eliminates many logistical problems. It has to be a key requirement.

How to Formalize the Statement and Make Sure Your Designers Stick to It

The six points previously listed are universal requirements for every manufacturer. Your particular design architecture statement will also contain items unique to your products or business strategy. The key point is to formalize this statement and require regular design reviews to ensure compliance.

List approved components. Some companies create a list of components approved for use. Deviations from this list are accepted only if the designer can prove to his peers that the design will be compromised by staying within the list's confines. This adds a self-policing activity within the design group. It also ensures that the list is changed to keep up with new technology. The architecture statement will stop many futile justification exercises comparing hard and soft costs. Establish the policy for designers to follow and check that they do.

Conduct design reviews. The design review is the place to ensure logistically friendly designs. It should be conducted with representatives from sales, product planning, design engineering, manufacturing, accounting, and purchasing. If the design process has been done correctly, most of these representatives will have been intimately involved with the design from its inception. They will have already provided their insights into the product.

The review simply permits the key business team members to agree formally with the design. It also ensures the design architecture statement has been followed. But the critical item the review must check is that the design will *not* be a failure because it ignored the key concepts of logistics.

HOW TO USE OVERPLANNING FOR CONTINGENCY TO FATTEN YOUR BOTTOM LINE

Let's turn to the hoist example in Figure 6-1 again. We stopped when we had planned all the modules using mix breakdowns based on historical or forecasted splits. The pendant was split 60/40 for a master schedule of 30 As and 20 Bs as shown in Figure 6-2. But what's the chance of the actual customers' orders splitting 60/40 in a given week? Not much.

And it's also true that the actual customers' orders in a given week won't match the other module percentages either. The chance of being able to ship 50 hoists in any week, with all the mismatched options, is zero. We'll be lucky to ship 40. Inventories will be high and customer service poor. Shortages will cause chaos.

Safety stocks of the various modules could offset the forecast error. Dipping into the safety stocks when necessary will allow matched sets of modules to be pulled and assembled. Fifty hoists per week of almost any configuration can now be assembled and shipped.

Another approach is to use overplanning. If we look at the pendant history again, we find that the average split was 60/40, but in some periods it was 70/30 and in others 50/50. Here we have the three numbers (expected, upper limit, and lower limit) discussed under forecasting. Based on this new information we are going to take the larger percentages, 70 and 50, and make the master schedule 35 of A and 25 of B, as shown in Figure 6-8, for a total of 60.

Many people react to this suggestion with "It will pyramid inventories at ten per week and demand capacity in excess of what is really needed." But this is not true. Without delving too deeply into the details of planning systems, they first apply any unconsumed inventory against needs before planning new production.

For example, let's assume we made 60 units of pendants this week. We will consume only 50 to be assembled on the actual orders booked, leaving a balance of 10. If we plan 60 again next week, the on-hand, unconsumed balance of 10 will be applied against the 60 leaving a net production quantity of 50. So all the 60 in the master schedule did for us is get 10 units of safety stock.

Product	Average Percentage Mix	Master Schedule Quantity	Overplanned Percentage	Overplanned Master Schedule
Pendant A	60	30	70	35
Pendant B	40	20	50	25
Totals	100	50	120	60

FIGURE 6-8. Overplanning the Pendant

Where capacity is concerned, as we plan to build only 50 hoists per week, we will ignore the extra 10 overplanning except for the first time we have to build it. After that, the allocation of inventory against the requirements will eliminate the need to have capacity for 60.

Overplanning—A Very Powerful Tool

Now let's take a look at the power of overplanning. If we put this extra 10 units into week one of the master schedule it will be the same as carrying 10 units of safety stock of the pendants. If we put it into week three, though, and keep the extra 10 always three weeks into the future by using a "time fence," then we will have only 10 sets of components extra as shown in Figure 6-9. We can now assemble the right 50 pendants when the orders come in if that can be done (along with assembling and testing the hoist) within our four-week D time. We carry less safety stock—only the components—but retain the same amount of flexibility in the marketplace. Similarly, if we put the overplan into week eight, and always keep it eight weeks into the future, then we would have only 10 extra sets of raw materials or long lead-time components. Putting the overplan and keeping it in week twenty would have no impact on the detailed planning whatsoever.

Standardizing the products as described earlier would make overplanning even more powerful. For example, the pendants before standardization would still need a high inventory of extra components to provide acceptable customer service. The reason is simply that the two buttons are the only standard items. All others are unique and so must be buffered from sales mix variations by excess inventory.

After standardization, when everything is common except the plug button and one switch, excellent customer service could be provided with minuscule excess inventories. This extra inventory could be provided through overplanning.

Achieving Both Mix and Volume Flexibility

So far we have talked only about mix overplanning. But it is obvious that if you add some reserve capacity in order to quickly expand flow rates, overplanning will also give volume flexibility. For example, if you have 10 sets of extra raw materials and long lead time parts in stock, put there with the overplan in week eight, you could increase the volume of output of pendants up to 60 at that time as long as you quickly increase flow rates through your affected resources by 20 percent. Of course you would need to do the same with all the other modules to get an increase in output of hoists.

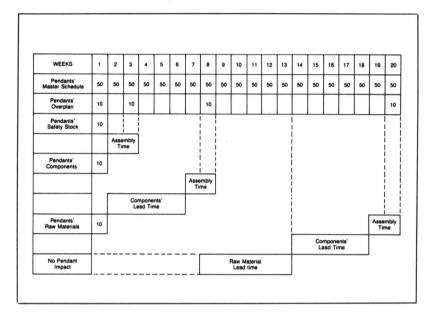

FIGURE 6-9. The Power of Overplanning

Why Overplanning Will Actually *Reduce* Inventories

Some people frown on the increased inventory investment overplanning requires. But that is not the point. The choice is either no overplanning or some overplanning. No overplanning will mean mismatched modules when customers order, huge inventories caused by the mismatched options and by shipping less than 50 hoists per week, poor customer service, and a disrupted factory. Overplanning will certainly generate some extra inventory; the amount will depend on where you put the overplan numbers in the master schedule. But if overplanning produces a better match between plans and actuals, then total inventories will be reduced. We'll ship the right 50 hoists per week, provide good customer service accordingly, and avoid factory disruption in the bargain.

Maybe the biggest benefit of overplanning, though, is in its role in getting top managers involved in the contingency planning concepts. The degree of flexibility you want in the marketplace must be traded off against inventory and excess capacity. This is a critical business decision. Don't leave it up to lower levels of management. Get some good business sense in this decision-making process.

PRODUCT DESIGN IS CRITICAL

The logistical characteristics of products are extremely important to their success in the marketplace. Too many products have failed, not because of their cost, functionality, or aesthetics, but because they were not available when the customer needed them.

Frequently, minor changes can convert a logistically unfriendly design into a very friendly one. The key, of course, is not to let the unfriendly ones out of the design department in the first place. Make the necessary changes while the products are still in the development stage. Don't let a potential world beater fail because somebody ignored its logistical characteristics.

7 | How to Work Your Plan Now That You've Planned Your Work

Planning the future is an essential activity. But as we've discussed more than once, no plan is worth making unless it is monitored and managed.

There are no operating plans more important to manage than the master schedule. If this plan is wrong, then huge commitments of company funds will be made to produce the wrong inventories. The master schedule monitoring process is often married to the technique of customer order delivery promising. In simple terms, the factory's future production, as defined in the master schedule, is consumed by orders booked. Promises for future orders are based on the amount of unconsumed master schedule in a given period. In essence you are simply selling the plan. If the volume of orders received deviates too far from the planned production rate, you need to change the plan.

TWO WAYS OF MAKING GOOD DELIVERY PROMISES

There are two fundamentally different ways of promising deliveries to customers. One is through a standard lead-time figure. The other is by comparing actual sales to planned production and promising out of the unsold planned production.

Many sales personnel like the idea of a consistent lead time. They don't necessarily want very short lead times. They prefer that quoted lead times be competitive and consistent; for example, a four-week lead time is always four weeks.

This is a difficult thing to actually do. Demand fluctuations are often more than the factory can absorb. It cannot adjust its flow rates quickly enough to keep demand and supply in balance. The only choice now is to adjust lead times unless additional buffers are available to absorb the demand variability.

When Can You Quote Standard Lead Times?

Standard lead times are possible only under some or all of the following conditions:

- The demand for specific products closely matches their actual supply.
- The total demand for all products matches the total supply, and mix variability is easily accommodated.
- The factory can flex its flow rates as quickly as demand fluctuates.
- Buffers, such as safety stock, can absorb the demand fluctuations and allow the factory to produce at an even rate.

Standard lead times demand that you closely monitor actual demand compared to supply. Imbalances in rate or mix must be quickly remedied to keep lead times stable.

The Logic Behind Available-to-Promise Deliveries

This technique starts out with a master schedule defining what products you intend to make, how many, and when. As orders came in, they are booked against this planned production. The unsold portion of planned production is therefore available to promise.

Available-to-promise logic can be used for any type of product, make-to-stock, assemble-to-order, or engineer-to-order. The difference is simply how you define "product" in the master schedule. This technique will give variable lead times to customers, controlled by the order demand rate relative to the master schedule's output rate. How this technique works can best be seen by analyzing two cases.

How a Textile Machinery Manufacturer Puts Available-to-Promise Logic to Work

Figure 7-1 shows a simplified view of the approach used by a textile machinery manufacturer. Their machines are used to treat fibers mechanically. They are equivalent to a machine tool or piece of equipment in any factory.

As you can imagine, this company makes a variety of different machine types. Within each basic family of machines there are many optional features customers can request. The permutations and combinations of possible finished goods customers can order runs into the thousands, certainly more than for the hoist example from chapter 6.

The company has broken its planning process into master sched-

BASIC MACHINE

Weeks	1	2	3	4	5	6	7	8	9	10	11
Master Schedule		50			50			60			60
Booked Orders		45			35			10			0
Available to Promise		5			15			50			60

OPTION A (60%)

Weeks	1	2	3	4	5	6	7	8	9	10	11
Master Schedule		35			35			42			42
Booked Orders		23			21			5			0
Available to Promise		12			14			37			42

OPTION B (10%)

Weeks	1	2	3	4	5	6	7	8	9	10	11
Master Schedule		6			6			8			8
Booked Orders		5			6			0			0
Available to Promise		1			0			8			8

FIGURE 7-1. Order Processing

ules for the basic machines plus separate master schedules for the options, not unlike the hoist example. The items in the master schedule are treated in isolation, as if the company could sell basic machines or options by themselves, when in reality they cannot.

The master schedule for each basic machine is based on the total demand for machines, a combination of booked orders plus forecasted business. This establishes the total rate of flow out of the factory.

When orders are received they are identified according to product family and slotted against the planned production. An available-to-promise line, meaning machines yet to be sold, is calculated in this fashion: master schedule minus booked orders equals available to promise. Figure 7-1 is for one machine type. The company has similar processes for all its other types of machines.

Option A is expected to be attached to the basic machine 60 percent of the time. This percentage could come from history or a sales prediction. You might think my mathematics is bad, as 60 percent of 50 machines is not 35 but 30. However, this company also uses mix overplanning, discussed in chapter 6, to improve its chances of always having option As on hand to go with basic machines. The company doesn't want to have basic machines (the high cost part of the product) available without options to go with them. If this happens, the basic machines sit in inventory waiting for options. Then the factory is forced to scramble to make options, inventories are high, and customers are not served well.

Any option A called out on the sales orders is booked against the planned production of option A, just as with the basic machine. An available to promise is calculated. Option B is handled the same way.

This simplified view shows only two options. In reality there are up to 100 different options per machine family and it's possible to have more than one option on any machine. The same logic applies, though, even with the larger number of options.

Order booking. Along comes an order for five machines all of which need option A. Delivery is requested as soon as possible. Both basic machines and option As are available in week 2, so the order can be promised then. The numbers are now recalculated to take this new order for five machines into account, as shown in Figure 7-2. Booked orders for basic machines are now 50, leaving zero available to promise. Option As are now 28 booked, leaving 7 available to promise. Option B was not affected.

Managing the overplan. Look at the three available to promises in week 2. No basic machines are available but options are. If the

BASIC MACHINE

Weeks	1	2	3	4	5	6	7	8	9	10	11
Master Schedule		50			50			60			60
Booked Orders		50			35			10			0
Available to Promise		0			15			50			60

OPTION A (60%)

Weeks	1	2	3	4	5	6	7	8	9	10	11
Master Schedule		35			35			42			42
Booked Orders		28			21			5			0
Available to Promise		7			14			37			42

OPTION B (10%)

Weeks	1	2	3	4	5	6	7	8	9	10	11
Master Schedule		6			6			8			8
Booked Orders		5			6			0			0
Available to Promise		1			0			8			8

FIGURE 7-2. Managing the Overplan

options cannot be sold separately for units already in the field, these available-to-promise figures in week 2 are useless. Worse than that, they are causing unnecessary production of all the ingredients that make up the options. The detailed logistics system underlying these numbers interprets these as real demands and is attempting to buy and make all the associated parts. Flow rate capability is being consumed needlessly, since these parts are no longer required. As a result, sequences are also wrong, and inventories of options are growing unnecessarily. It's obvious, therefore, that the available-to-promise quantities in week 2 for options A and B need changing, but how?

Two key choices. There are only two things you can do with the unwanted overplan. Remove these numbers completely (in other words, drop them), or push these numbers to a later date to coincide with the next available-to-promise figures for basic machines.

For option A, the best choice is drop the number completely. In week 5, the next time basic machines are available, we already have 14 option As in the plan and only 15 basic machines. That is almost 100 percent coverage for a 60 percent option. It would be pointless to increase the 14 option As to 21 using the 7 extra from week 2.

For option B, the best choice is to defer the extra unit available to promise from week 2 to week 5. All planned option Bs have been sold already in week 5, but there are still 15 basic machines unsold. The option B in week 2 should be pushed to week 5 and consideration also given to increasing the number of option Bs produced according to the master schedule in week 5.

Clarifying the Master Scheduling Concepts

This example shows various master schedule concepts clearly. The options have been separated out to improve forecasting and to allow overplanning for mix variability. This company cannot afford to volume overplan because that would mean overplanning basic machines. Their stacked lead time is seven months and the basic machine contains some very expensive long lead items. Not only that, the factory's flow rate is not easily adjusted. Hence the extra costs to flex volume are not worth the incremental sales.

They are being stable with the factory and vendor schedules but flexing to the marketplace, at least for the options mix. They do cause some instability when they delete or defer any unwanted overplan as detailed schedules for the option components will need rescheduling. The detailed logistics process will read the changed overplan as reduced or deferred demand and now delay buying and making all associated parts. But this company says they can manage deferrals of production or vendor deliveries much more effectively than expediting. They apply a reasonability test to the delays and implement only those with a large capacity or inventory effect.

What Are the Results?

This process gives excellent results to this company. It actually ships 95 percent to 98 percent of its machines on time to the delivery dates that are promised. Lest you think the company does not aggressively promise short delivery dates, think again. It tries very hard to accommodate its customers' needs. But the company refuses to lie. It will only promise a delivery when it is very sure it can live up to its promises. Total inventories are low. The only excess the company has are some parts caused by the unwanted overplanning. Costs are reasonable because of the stable factory and vendor schedules.

How a Motor Manufacturer Keeps Its Available-to-Promise Line Flexible

Another example of the master scheduling process, similar to the textile machinery company, is shown in Figure 7-3. This example comes from a company that makes specialized medium horsepower electrical motors. No two motors are the same, as they are all engineered to unique customer requirements.

This company's problem is that its P time is 18 weeks. Customers won't wait that long—they want delivery in 6 weeks. So this company must complete 12 weeks of its P time before a customer's order arrives. Fortunately, this is the procurement segment of its P time, leaving 6 weeks for machining, assembly, test, and shipping.

The engineers have agreed to design each family of motors using

FAMILY 1

Weeks	1	2	3	4	5	6	7	8	9	10
Master Schedule	30	30	30	30	30	35	35	35	35	35
Booked Orders	28	27	29	34	29	36	33	25	10	2
Overplan							35			
Available to Promise	2	3	1	−4	1	−1	2	10	25	33

FAMILY 2

Weeks	1	2	3	4	5	6	7	8	9	10
Master Schedule	100	100	100	100	100	100	100	100	110	110
Booked Orders	103	105	101	97	103	92	90	81	28	3
Overplan							100			
Available to Promise	−3	−5	−1	3	−3	8	10	19	82	107

FIGURE 7-3. Family Master Schedule

a limited selection of raw materials, including standard castings for the body, bar stock for the shaft, steel sheet for the laminations, copper wire for the windings, and so on. These purchased items are listed in a family bill of material with their expected rate of usage per family unit.

The master schedule quantities in Figure 7-3 drive this family bill of material to provide raw materials. As customer orders are received, these motors are engineered and booked as specific products. Their production consumes the raw materials provided under the master schedule.

However, the actual sales by product family could be different from the master schedule numbers. It's entirely possible that the company might book more of one family and less of another. To provide a buffer, an overplan quantity is strategically placed in week 7. The overplan quantity drives the family bill of material which gives the company additional raw materials. These overplan quantities allow the mix to fluctuate to some degree between families without any problems. (Figure 6-9 shows how the overplan provides for sufficient raw materials.)

This overplan is also a volume overplan. The general manager insists that the company produce the master schedule numbers working a straight 40-hour week. Overtime is reserved to adjust flow rates quickly when there is a need to change volumes to grab opportunity business.

In this case the overplan is a "rolling" overplan. It never moves inside the six-week time fence signified by the double vertical line. The system's logic keeps these numbers always in the seventh week, so they generate only excess raw materials.

The available-to-promise line is not as exacting in this example as in the earlier textile machines example. As long as the total demands on the business do not exceed its flow rate capability, the company lets each week's booked orders in each family vary to some degree from the master schedule quantity. Only when the total demands over some period exceed the master schedule does the company turn on the overtime.

The outstanding results. What are the results? The shipment performance-to-promise date is between 98 and 100 percent. Total inventories have dropped significantly since the company started this process. Factory and vendor schedules are reasonably stable. The company grabs market share from the competition at every opportunity with its volume flexibility. Lead times rarely vary very far from their objective of six weeks.

All Products Need a Similar Process

The two previous examples are for an assemble-to-order product (textile machinery) and an engineer-to-order product (medium duty motors). A make-to-stock product needs the same kind of monitoring of demand and supply. Its specific monitoring methodology is not quite the same, but the concepts are identical.

In all cases the objective is to compare actual demand to planned supply. Deviations must be quickly highlighted for resolution if you want to manage the operations well. Don't forget your limited options—change customer lead times, use buffers, have a flexible factory, manage the demand, or some combination of all the above. You must make this choice for your company.

STUFF IT AT YOUR PERIL

I emphasized the need for flow rate balance in chapter 2. The master schedule must be realistic and feasible based on the actual flow rate capability of upstream resources. But many managers are unwilling to demand this balance. Comments such as, "let's put the work in the factory and see what drops out," or "let's task the factory to produce," show a clear lack of management understanding of the consequences. Never refusing an order based on delivery lead time is also a killer.

Here's what happens when the master schedule's desired rate of flow exceeds the upstream resources capability. I'll use a simple example to show the problem. Four fabrication work centers produce parts for one assembly department. The assembly work center also packs and ships the end products.

The master schedule's desired rate of flow is equivalent to 120 hours per week of production in each of the fabrication resources. All of these resources have an actual production capability of 100 hours per week.

This means all four workcenter supervisors have a sequence list of items to produce this week equivalent to 120 hours of work. They all have a capability of 100 hours per week as just stated. Now what?

The choices are few. First, you want all upstream resources to increase their actual flow rate capability to 120 hours per week quickly.

Overtime, subcontracting, and hiring are all potential ways to improve output.

But what if that can't be done? Maybe they're already working overtime just to produce their 100 hours per week. Getting qualified subcontractors may be impossible because of a proprietary process (that is, the process cannot be revealed to subcontractors, or when specialized machines are needed that typically subcontractors don't have). Hiring skilled people is usually a long-term, not a short-term, proposition.

The Disastrous Consequences of Ignoring Flow Rate Balance

The default choice is this: if the actual flow rate capabilities are 100 hours this week but the sequence list of items to produce this week adds up to 120 hours, then each supervisor must choose which items to make. A variety of methods can be used to help him make the choice, such as hot lists (an informal priority system) shortage meetings (when we are out of something we need it's an obvious priority) and gravy jobs (these are the ones that make the supervisors look good).

Every supervisor in the plant is making individual choices and is being "helped" by one, two, or all three of the methods just mentioned. What chance do the supervisors have of making all the matched sets of parts needed in the assembly department or to suit the real needs in the marketplace? The answer: slim.

So 100 hours worth of items are being made by each fabrication work center but because they are not matched sets, the assembly department cannot assemble and ship products. They have shortages of some items and plenty of others, but nothing that matches.

The assembly department now gets into the expediting business and tries manually to give correct sequences to the fabrication work centers. This may be possible for very simple products, but people cannot do this successfully with 10,000 different parts, complex routings, and a dynamic factory. Even so, they still try. To show you how ridiculous it can be, I once visited a company that had transferred 16 skilled assemblers to the expediting department to expedite fabrication work centers that didn't have enough flow rate capability to support the master schedule. You can imagine how much that "helped" the situation.

Returning to my simple example, four fabrication work centers are each producing 100 hours per week's worth of unmatched parts.

The assembly department wastes a lot of time working around the shortages so its productivity is low. Inventories of the unmatched parts are high and growing. Actual output of the factory is probably around 80 hours worth of product per week. Customer service is poor and deteriorating.

WARNING: THERE'S ONLY ONE RIGHT WAY TO PRESSURE FOR MORE OUTPUT

Many people feel factories produce more under pressure. Factories have helped foster this belief because they often "pull rabbits out of the hat" at the end of the month, end of the quarter, or end of the year.

I too believe that factories produce more under pressure. What is critically important, however, is the method of challenging the factory to do better. Pushing for more productivity, less idle time, and less rework, is the right approach.

But doing the pushing through excessive numbers in the master schedule is the wrong approach. As my simple scenario of the four supervisors demonstrates, overstating the master schedule makes total output drop, not increase, because of the "cherry picking" that is now forced to occur. The only reason that output increases at the end of the month, quarter, or year is that people manually figure out what is possible to ship and concentrate all their resources on these limited items. The formal system, driven by the overstated master schedule, is ignored, as it must be. Supervisors and schedulers manually make realistic plans and then flow the right amount of the right things through all resources to match them.

This is obviously the correct way to work all the time, and by doing this, you will pull rabbits out of the hat every day. But this means driving the formal system through a reasonable, capacity-sensitive plan. And one of the prime tasks of the general manager is to make sure this is done.

Gauge the Problem In Your Own Company

To get some idea of the problem in your company, ask your assembly supervisors, preferably after they have had a couple of strong drinks, "How much more product could your people assemble if you had matched sets of parts?" The answer in most cases will be, "At least 20 percent," and you know they are sandbagging. One company documented a 30 percent increase in output when it balanced the master schedule to upstream resource capability and hence developed synchronized parts production.

I used internal work centers in my simple example, but the same scenario holds true for vendors. They have a limited amount of flow rate capability assigned to your business. Force them to exceed this capability by demanding too many items and your vendors will make choices to suit themselves. Again, unmatched parts will result. Check this situation by asking the store's personnel if they have excess inventories of some things and shortages of others. If the answer is "yes," and the imbalance is severe, you know that unsynchronized choices are being made.

The example I've used is oversimplified. Consider the real-life complexity of many plants with hundreds of vendors and maybe 50 or 60 work centers. Keeping them all producing the right thing at the right time is not easy. It's impossible with an overstated master schedule.

THE PERFECT MASTER SCHEDULE

A perfect master schedule is one the factory *exceeds* regularly by a very small margin. A poor one is a plan the factory fails to produce by any margin. This is entirely contrary to many people's thinking. They believe you have to overload a factory to get it to produce at the desired rate. Their reasoning is that a factory is a complex place. Something is always going wrong. If you need 100 units per week out of the factory you have to ask for 125. This way you have something in the schedule you can pull up if you encounter problems, and you can still end up shipping 100 units per week.

I hope my simple example persuades you *not* to pressure the factory through the master schedule. Overstating it will *guarantee* something goes wrong. Supervisors or vendors will make choices that don't end up as matched sets of parts.

If you wish to pressure a factory to produce more, do it through measures of productivity, standard hours generated, shipments, or some other measure. Don't do it through the master schedule. If you are concerned about having something extra in the plan you can pull up if needed, do it through volume overplanning. But make sure you realize capacity must be added if you try to execute this volume overplan.

Don't forget, *an excellent master schedule is one the factory exceeds regularly by a small margin.* There's no risk of the factory running out of work with some strategically placed overplanning. You can release this into the factory to fill any voids. What you guarantee is matched sets, the only production worth having.

TOP MANAGEMENT'S ROLE IS CRITICAL TO RUNNING THE BUSINESS RIGHT

Master schedules are key sets of numbers that commit large amounts of company resources to specific things. Hence they are key controllers of the operations side of industry.

Top management must take charge of these numbers. Too much business strategy is at stake for these numbers to be left to lower levels of management. Senior managers must understand how these numbers are used and where they fit in the hierarchy of planning.

Some policies and guidelines will be needed to administer the master schedule. For example, how should products be designed to assist the forecasting process? How much flexibility must be provided in the actual output capability, in terms of both mix and volume? Who is authorized to make changes to the master schedule? (This could be a variety of people, depending on the magnitude of the change.) What do we do when demand exceeds supply by a certain amount? How will we manage the demand side to keep it in step with supply? These policies and guidelines may be suggested by lower level managers, but they must be reviewed and agreed to by top management to ensure the business is run according to their wishes.

Top Management's Role in Resolving Limited Objectives

Conflicts are inevitable in business because the various organizational functions in industry are charged with limited objectives and measured on attaining them. Only top management can resolve these conflicts.

Figure 7-4 shows these conflicts in simplistic terms. If you look closely, you'll find that every objective underneath each function's title conflicts with another function's objectives. For example, flexibility under sales conflicts with stability under manufacturing. Product variety under sales conflicts with both investment and cash flow under finance.

The objective is to develop a master schedule that simultaneously achieves all these goals. The words "compromise" or "tradeoff" shouldn't be used until you have exhausted all your options to achieve this seemingly impossible task. Compromise means you accept fail-

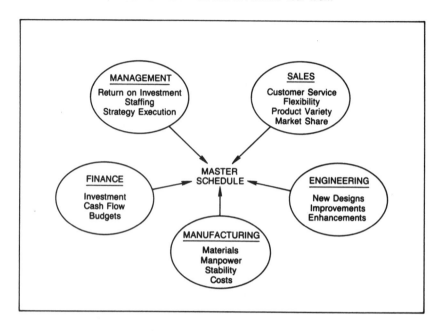

FIGURE 7-4. Functional Conflicts

ure. You have two objectives and cannot achieve both, so you settle for something in between. For example, your wife wants to go to the beach for vacation, but you want to go to the mountains. So, as a compromise you stay home. Or, in the case of my family, I go to the beach with my wife. I know who's boss!

Achieving all of the functional objectives 100 percent *is* an impossible task. Your skill in creating the master schedule and implementing the strategy around it will determine how closely you can come. Then, you need only resolve the remaining conflicts. Only top management can do this most effectively because it is looking to optimize the business in total.

WHAT'S INVOLVED IN MONITORING THE PLAN

The master schedule must be watched and measured aggressively. Its key role demands that significant deviations be quickly highlighted for resolution.

Most companies review the master schedule on a monthly cycle.

MANAGEMENT SUMMARY
AS OF FEBRUARY 28

MONTH		JAN.		FEB.		MAR.	
PERFORMANCE		PLAN	ACT	PLAN	ACT	PLAN	ACT
Family #1 (Units)	Sales	57	60	57	56	57	
	Production	55	54	56	58	56	
Family #10 (Units)	Sales	13	10	13	12	13	
	Production	12	13	12	12	12	
Repair Parts ($000)	Sales	25	23	25	28	25	
	Production	25	26	25	25	25	
Beginning FGI	($)	1153	1153	1110	1085	1083	1091
Production (MPS)	($)	1038	1025	1074	1095	1037	
Cost of Sales	($)	1081	1093	1101	1089	1121	
Sales Forecast	($)	2162	2186	2202	2178	2242	

FIGURE 7-5. Master Schedule

The data are summarized into a few useful families, and reports are generated similar to that in Figure 7-5. A couple of months of historical performance plus projections of inventories, production rates, cost of goods sold, and net sales allow top managers to see quickly whether the plan is reasonable. They can see where the problems exist, and whether they are sales or production problems. The status of new designs and important modifications to old ones should also be reviewed in the master schedule meeting as well as the status of any significant change in resources, such as subcontracting or hiring.

KEEP THE HEAT ON BY ASSIGNING CLEAR-CUT RESPONSIBILITIES

The company depicted in Figures 7-1 and 7-2 attempts to clearly delineate responsibility for executing the master schedule. Their approach is that manufacturing's job is to adhere to it as closely to 100 percent as possible. Manufacturing shouldn't concern themselves with the sales of the products. That job belongs to sales, which should sell the master schedule as it was built around their forecast. That way, if machines come down the assembly line without any customer orders, the company puts heavy pressure on the sales force to sell them. If sales fails to respond, the company puts a limited number of basic machines in stock, maybe assembled with frequently ordered options. These machines are kept in the sales department's inventory. The pressure for inventory reduction keeps the heat on sales to sell these machines as soon as possible.

In actual practice, responsibilities are not this clear-cut. The general manager adheres to this principle as far as possible, however. He considers the master schedule a contract between sales and manufacturing: sales' job is to sell it, manufacturing's job is to make it.

WHERE SHOULD THE MASTER SCHEDULER REPORT?

The master schedule needs careful thought in its preparation and some management on a day-to-day basis. Various alternatives must be developed to accommodate changing business conditions. Senior management must review the strengths and weaknesses of each alternative and make decisions accordingly.

The person or persons charged with responsibility for the master schedule must have a broad view of the business, without bias toward any one functional group. This is not the case with most master schedulers today, however. All too often, they are buried in the depths of the materials department, which traditionally reports to manufacturing.

In a few enlightened companies, however, master scheduling is a peer function side-by-side with the heads of finance, engineering, sales, and manufacturing, all of whom, in turn, report to the general manager. That makes the master scheduler's job a senior level position, separate and distinct from any functional group.

I predict this is where this function will end up for most companies eventually, as they come to recognize the power of the master schedule numbers. To round out this position, the master scheduler will also be involved in monitoring the strategic business plan and other long-range planning activities.

THE MASTER SCHEDULE IS KEY

Chapters 5, 6, and 7 have defined the master schedule, why it exists, and how to manage it. Numbers in the master schedule are among the most vital for every company.

Every company has a master schedule, by definition. The only question is who is involved in its development, and at what level. Keep in mind that huge commitments of company resources are made daily to buy and make specific items to execute the master schedule. The better these numbers, the better the execution.

The P:D ratio is a key concept that must be carefully evaluated for every product. Your choices for solving this planning dilemma are limited, so pick the ones that best suit your business.

Overloading the master schedule is not the answer to increase production. In fact, it has negative effects. Overloading is an ostrich-like reaction for avoiding tough decisions. Make sure you understand these negative effects and insist on a realistic plan, one you are sure the factory and vendors can execute.

Top managers must consider the master schedule to be their handle on the operations side of the business. Policies, guidelines, and above all else, master schedule review and management are essential in order to get effective results.

8 | Slash Business Risk and Improve Customer Service by Reducing Stacked Lead Time

The P time for a product was defined in chapter 3 as the stacked or aggregate lead time required to procure raw materials and produce an item. In some cases, it also includes engineering or product specification before procurement can begin. In other cases, it also includes distribution of the product from the factory gates to the end customer. Figure 3-1 shows the P time for a product, but in this example it includes only procurement and production time.

The discussion from chapter 3 on the P:D ratio defined utopia as P less than D—that is, a stacked lead time shorter than D, the customer lead time. P less than D eliminates speculation and risk on your part. The best way of improving the P:D ratio is by reducing P, not increasing D. Increasing D rarely eliminates speculation; instead it transfers it to the customer.

The full impact of P less than D on planning and managing, (see chapters 5 and 7) is enormous. P less than D means that no specific materials, products, intermediates, subassemblies, and so forth are procured or produced based on speculation. It means every item of inventory in a company is there to support a known customer's demand. In other words, you are now making wholly to order, buying and making nothing to stock. This should be the strategic objective of every manufacturer.

If your P is less than D, you need not concern yourself with standardization, simplifying the product line and forcing a match, except insofar as these strategies help to further reduce your P time. You don't have to concern yourself with safety stock and mix over-planning, either.

Even if your P is less than D, however, you still need to forecast the future. But now, you are making forecasts for capacity and flexibility reasons, not for scheduling specific items. As long as your manufacturing processes are flexible enough to accommodate specifics, you can rely on aggregate forecasts for capacity planning. And as we all know, forecasts for aggregates are more accurate than for specifics.

THE ELEMENTS OF LEAD TIME

Lead time (P time) is a complex subject because there are many different kinds of lead times. Procurement, vendor, manufacturing, engineering, tooling, and customer are only some of them. Planned lead times are what you expected lead times to be and therefore instructed the logistics system to use—versus actual lead times, that is, what it really did take to do. You can see the confusion. An item with a planned lead time of six weeks can be made overnight if it's urgent. So, what is lead time really?

We will dissect it as shown in Figure 8-1. First, we'll focus on the manufacturing lead time to establish some concepts and define the various elements. We will then explore programs to reduce lead times and their synergistic implications.

The ideas I will discuss all come from the Just-in-time (JIT) philosophy. One of the key objectives of this approach is to reduce lead times to avoid the associated NVAW (Non-Value Added Waste) long lead times cause.

I will use as a case study the production of a batch of machined components needed to assemble end products. Let's assume the raw materials are currently in stock, so we can exclude procurement time. We have made these items before, so we have in hand the engineering design, process instructions, and any necessary tooling. Our only question is what various activities must occur that, when added together, make up the planned lead time for this batch of components.

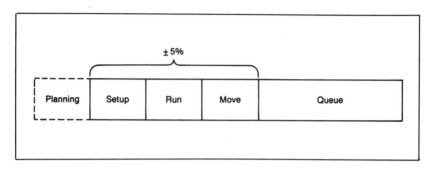

FIGURE 8-1. Elements of Lead Time

We'll take a look at each of the individual elements that make up lead time. Before I go any further, let me make sure our objective is crystal clear. What we want are shorter and shorter P times. This means that every element that makes up lead time must be reduced —and not necessarily along the critical path. It could be that the critical path items are standard to large product families; hence their planning is very reliable (see the mushroom, Figure 3-7). The non-critical path items could be the ones that add variability. In that case, it would be much more important to reduce *their* lead times rather than those along the critical path.

The Planning Element

As shown in Figure 8-1, the first element of lead time is planning. Before a batch of parts can be made, some planning or preparation activities must take place. These activities could include determining the need for parts, writing shop orders or schedules, and planning their production. The broken line around this box signifies that these elements should not be included in the calculation of P time. These activities can be done in tandem with other logistics activities, such as purchasing the raw materials. Planning time could also include picking the material and delivering it to the first operator.

The Setup Element

The next element of lead time is set up. Figure 8-2 shows a comparison of setup times for large stamping presses in three automotive companies. Setup time in this context is the time from when a machine or process stopped making the last good piece of a batch of product A and starts to make the first good piece of a batch of product B. This

Company	General Motors	Volvo	Toyota
Setup Time	4 hours	2 hours	10 minutes (SMED)[1]
Average Batch Size	20 days' supply	10 days' supply	1 day's supply
Current S/U Objective	10 minutes	10 minutes	100 seconds (OTED)[2]
Future S/U Objective	?	?	zero (NTED)[3]

[1]SMED = Single Digit Minute Exchange of Dies
[2]OTED = One Touch Exchange of Dies
[3]NTED = No Touch Exchange of Dies

FIGURE 8-2. Setup Times Comparison

is more correctly called "changeover" time, and I will use this term in the rest of the discussion.

Several things jump out at me when I look at Figure 8-2.

Changeovers. First, assume we have 20 changeovers along a product's logistics chain. Each changeover originally took 4 hours, but let's say we can cut that time to only 10 minutes. We save less than 5 days, assuming a two-shift operation. With P times for GM of 11 weeks, (see figure 3-6) what's the big deal?

The answer? Not much, directly. But indirectly, long changeovers are to blame for a large part of the P time, much more than the direct saving of 5 days would indicate. As a matter of fact, changeover reduction is one of the key cornerstones of the JIT approach as we shall see later.

Inventory. Second, take a look at the enormous difference in inventory levels among the three companies. Where does GM store 20 days' supply of hood lids, fenders, trunk lids, and so forth? Obviously, in automated storehouses, or in transit between various plants or in work-in-process. But these storage and transit costs are not adding value. Neither is the inventory, produced efficiently but now lying idle—an excellent example of NVAW.

Flexibility. Third, how much more flexible at no cost penalty is Toyota compared to GM or Volvo? Toyota can produce any needed item very quickly. What may not be so obvious is that consumption of items during the final finishing process is usually one at at time.

If you produce in large batches upstream, the chances of having all the parts you need when you need them downstream is low. Shortages are almost foreordained. But if you produce upstream in small batches very often, you greatly enhance your ability to have matched sets of parts.

HOW TO REDUCE CHANGEOVER TIMES

What is required to reduce changeover times? To a large part, nothing more than the determination to do so. An evaluation of the changeover process in most plants shows that it has received very little attention. We assign this activity to our most skilled people and assume they know how best to do the job. Nothing could be further from the truth.

Three Ways to Improve Changeover Times

Three basic categories of changeover improvement are possible with a little understanding.

1. *Eliminate external time.* The changeover process can be split between external time (time when the machine or process is traditionally stopped but could be running), and internal time (time when the machine or process must be stopped to accomplish the changeover).

Examples of external time include bringing in the tools needed to make the next part, scheduling the setup people to be at the machine when it completes its previous run, and bringing in all special clamps, bolts, hand tools, and gauges so that they are ready to accomplish the changeover.

With better preplanning before the change and a more organized way of clearing the area after the change, you can typically achieve a 50 percent reduction in changeover time. This can be done by simply keeping the machine running while the external activities are performed in parallel. A better way is to organize the area around the machine so all the needed tools, gauges, and handling aids are always there. This improves the setup people's productivity at the same time it reduces the changeover time.

2. *Streamline changeover methodology.* The method used to perform the changeover as mentioned earlier is usually left to the skilled setup people. But a standard industrial engineering analysis of the methods used can often reduce the internal amount of time by 50 percent or more. You can save time by using two people rather than one, eliminating unnecessary lost motion, clamping only where necessary, or using quick clamp devices wherever possible.

3. Eliminate adjustments. Many tools and machines need careful adjustment before they can make a good part. This is often accompanied by making trial parts, inspecting them, and adjusting several times before a good part is made. By doing this, you not only lose machine time, you also waste raw materials.

An evaluation of the tooling and the machines can often show where the adjustment can be stopped altogether, using standard tool geometries and scheduling them only on certain machines. Doing the adjustment off line on a special jig (in other words, converting internal time to external time) can also reduce adjustment time. A 30 percent reduction in internal time is typically achieved.

How One Company Achieved an Astounding 83 Percent Changeover Reduction

An American company applied these principles to the changeover of a stamping press. The results are shown in Figure 8-3. A video camera and recorder were taken onto the shop floor and an actual changeover

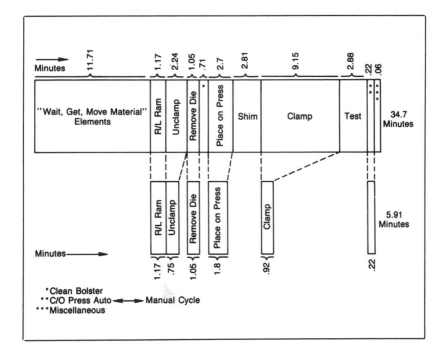

FIGURE 8-3. Setup Reduction

before improvement was filmed. The setup people, an industrial engineer, tool engineer, and scheduler, then went to a conference room, where they defined and timed each element of the change. After some brainstorming to improve the method and making some modifications to the tool, they succeeded in achieving an 83 percent reduction in changeover time. The out-of-pocket costs to achieve this were minimal.

When they started out, their objective was to achieve Single-digit Minute Exchange of Dies or SMED, which means changeovers in less than 10 minutes. The term SMED, along with OTED (One Touch Exchange of Dies) and NTED (No Touch Exchange of Dies), comes from Toyota. They are Toyota's method of focusing attention on the changeover process. Toyota sets these goals and then measures to achieving them. As one goal is passed, the next is set. Toyota's terms are now universally accepted as the targets for all manufacturers to achieve.

Shigeo Shingo in *SMED—A Revolution in Manufacturing*, Productivity Press, 1985, describes this process clearly. He demonstrates how to reduce changeover times on almost any machine with some creative thought and dedicated people.

The American company in our example has achieved SMED.

To get to the next level, One Touch Exchange of Dies (OTED), which is changeovers in less than 100 seconds, will require more thought, probably more tooling modification, and more practice in making quick changes. And it will probably cost more money. But the cash is readily available, freed up out of the inventory that has reduced so remarkably in tandem with the changeover reduction.

How to Eliminate Changeover Time Entirely

No Touch Exchange of Dies (NTED), that is, changeovers in zero time, may not be possible for this process or could be very expensive. But many processes and machines are already at the NTED condition, a utopia that *can* be reached. One way to achieve NTED is to buy or make several small machines that make only one product and never need to be changed over, rather than buying one large machine to make several parts. Another way is to buy flexible machining systems, computer controlled, to change from one item to another instantaneously.

It is obvious that with NTED, the quantity of an item that is economical to produce is one. And this one can be produced any time it is needed, not be buried in a large batch, produced in the hope it will be needed.

The Western World Missed an Opportunity

Why have we not addressed the problem of changeovers in the Western world? I believe there are three basic reasons. First, we have huge space, so storing things is no problem. In America, 30 to 40 percent of prime manufacturing space is used to store inventory.

Second, the calculations for Economic Order Quantity (EOQ), developed in 1915, say in essence that changeover time is not a problem. Supposedly, producing a large batch negates the negative influence of long changeovers. In retrospect, it seems a little incongruous that we accepted the idea of building a large idle asset (inventory) because we had a large idle asset (machinery).

Third, our financial performance systems don't focus attention on the problem. They hide it by either putting the time spent on changeovers into indirect accounts, so that it shows up as overhead, or apportioning it over the batch quantity so its impact on unit costs is low.

Ask any industrial engineer, "In your career, how much time have you spent either learning about improved changeover times or implementing faster changeovers compared to the time you have spent learning about or improving the speed of the process after it has been changed?" The invariable answer is, "Almost zero on changeovers, because it has such a small effect on unit costs and almost 100 percent on improving the process because that saves money!"

I gave a talk to the senior managers of a division of a large European company several years ago. The subject of changeover reduction came up as a method of reducing P times and increasing flexibility.

The vice-president of automation, the man responsible for the group that designs and builds the division's specialized production machines, told me, "In my 35 years in this division, I have never seen a specification for the time the machines should take to changeover from one product to the next. I get specifications of the product ranges to be built, rates of production, quality levels, and costs, but never once changeover time. I solve the unit cost portion of the changeover time by insisting that we always run large batches."

Is it any wonder, then, that changeovers on the machines his people design take several hours in most cases? The batch inventories this causes are a direct result of this lack of attention.

Warning: Be Careful of Your Reward System

There is a terrible pitfall awaiting the unwary in all this discussion of changeover reduction. It is the performance measurement system of the factory people. They are traditionally measured on efficiency, indirect/direct labor ratios, machine utilization, and so forth. Rarely are they measured on inventory levels.

Faced with this kind of reward/penalty system, it is easy to see how factory people could be tempted to reduce the changeovers as described but maintain the batch quantities high. They would get a better scorecard this way.

You must head off this problem by demanding smaller batches commensurate with the reduction in changeovers. If the machine is a bottleneck machine, slowing down the output of the plant, then using some of the changeover reduction to increase output and solve the bottleneck problem is obviously valid. But on nonbottleneck machines, it is mandatory that changeover reductions result in batch reductions in proportion. Shorter P times and inventory reductions are far more important than making a nonbottleneck machine more efficient so that it has more idle time.

This dilemma of traditional measurement systems stimulating poor business actions is one that will recur several times in this book. I feel so strongly about it that I have devoted all of chapter 11 just to this subject and how to counteract it.

You may wonder why I have spent so much time on the changeover portion of lead time when in total it is a small part. It is because of the synergistic effect that reducing changeovers has on the other elements, as we will see in later sections.

REDUCING RUN TIME

Run time is another element that makes up lead time. Most production, at least in the early stages of the logistics chain, is done in batches. These batches then move serially through their various processing steps. Accordingly, the run time to include in your lead-time calculation is the run or actual processing time for a batch. Reducing changeover times allows a reduction in the batch quantity, hence a proportional reduction in run times. Synergy is already at work.

HOW TO SLASH LEAD TIME BY CUTTING MOVE TIMES

Move times are yet another element of lead time. These are the times needed to move products from one process to the next. This may be between processes within the factory, between factories, or between the factory and various subcontracted processes performed by vendors.

A good analogy for factory move times is road traffic. It is obvious that to move quickly and reliably between two points, A and B, by automobile, utopia would be to have points A and B close together with a wide, straight highway between them. The worst scenario would be to have points A and B far apart with a city between them and no way to avoid the city streets.

What do you have between your receiving dock and the shipping dock, highways or a downtown? If you agree that the mission of manufacturing, as stated earlier, is to get the right materials and move them as quickly as possible from receiving to shipping, transforming them along the way, you obviously should have highways.

Tour Your Plant to Find Your Traffic Jams

Take a plant tour, as I do when visiting a company for the first time. Ask your tour guide to start where the raw materials arrive and then walk you through the various processes, leaving none out, until you get to the shipping dock for finished products.

On these walks I am never sure whether my tour guide is lost, drunk, or having a huge joke at my expense. We move back and forth between the various processes and departments until it seems more by luck than good management that products ever find their way to shipping. I sometimes think the Bermuda triangle exists in many plants.

By extending this thinking to your total logistics chain, you can uncover all kinds of traffic problems. For example, a large computer manufacturer has 29 factories involved in making one of its major products with 87 resource/product interfaces. How's that for a downtown to deal with?

Returning to the movement of materials within the factory, it is obvious that the layout of the various processes is the key to short move times. Functional groupings of machines, where technical specialties are kept together, always create confused product flows. Breaking these functional groupings apart into product groupings, sometimes called cells, where all the machines necessary to make a given family of items are located together and in the correct production sequence, will create highways for these items. Move time, because of the close grouping, will be zero or close to it. Combine this with performing more operations on one machine to avoid the moves altogether and you can reduce move time even further.

Simplifying the Environment: A Crucial Step in the Race Toward the Automated Factory

The management of "highway" groups is much easier than "downtown" groups. Scheduling, tracking, and performance reporting are all easier, leading to far simpler control systems. Instead of using complex systems to manage complexity, we have moved to simple systems managing simplicity.

This drive for simplicity of environment is crucial in the race toward the automated factory. We will discuss this in chapter 12, but it is obvious that the simpler the environment, the easier and quicker it can be automated. As this race is the next big challenge for manufacturers worldwide, preparing for it now and getting the right conditions for automation in place obviously give you a head start.

Some companies will find this idea of product or process groupings difficult to create. Their comment, "We have a job shop," implies that all products are different. The focus shouldn't be on variety of product, however, but on similarity of process. The products can be infinitely varied, as they are on an automobile assembly line. But if the processes are similar, then process groupings are possible.

Group Technology (G.T.) is a method of classifying products so their similarities can be identified. This is one way to identify the products that have similar processing steps and so could be produced in a cell. Another way to identify groups is simply to sample some products you currently have in production. If they fall into fairly well-defined process families, then this concept of cells has huge potential for your business.

QUEUES: OVERCOMING THE BIGGEST OBSTACLE TO SHORT LEAD TIMES

The queue is the last and longest element that makes up lead time. As you can see from the percentage of lead time typically consumed by setup, run and move, shown in Figure 8-1, the figure is not drawn to scale. It can't be, because the page isn't big enough!

Queue occurs simply because there are always jobs waiting to be processed by a resource. When a new job shows up, it has to wait for all the preceding jobs to be processed before it can be worked on. Queue is therefore proportional to the amount of work-in-process. If you want to reduce queue, by definition you also want to reduce work-in-process inventory.

We've already discussed the linkage between inventories and P times. I hope it is clear that these are in fact proportional.

For example, if you had zero work-in-process inventory, and machines and people idle waiting for work, then if one order to make a batch of products was released into the factory, it would flow through that factory very fast. Now if setups were reduced to zero, an order made for only one piece, and our factory organized into a series of highways, then this order would flow through the factory in the actual processing time of the product—less than a few hours in the majority of cases.

This is exactly what expediters do to the "hot" jobs in a plant. They leap long queues in a single bound! They force the processes to be made available as needed, and their one hot job is served regardless of the existing queue. They also split lots to get small batches of a variety of parts flowing to the finishing operations.

This is analogous to an ambulance on a crowded highway. With much noise and added confusion, drivers pull over to the side to let the emergency vehicle through. It makes fair progress, but not without some increased risk to the occupants of the emergency vehicle and the regular drivers.

The trick is to get all vehicles (jobs) moving quickly to their destination. This can be done only when a highway (factory) is not crowded. Then everyone, including the emergency vehicle, travels quickly and reliably to their destination. And the emergency vehicle travels at a much faster and safer rate than when it is forcing its way down the crowded road (queue).

The full implication of queue on P time can be seen in Figure 8-4. Products sitting idle waiting to be processed make up the major portion of P time. This also shows you that products rarely flow through a factory as the refinery analogy suggested. They lurch through a factory, instead. Operations are performed very efficiently on high speed machines for a short period of time, and then the work sits in queue for hours, days, or even weeks, waiting for its next chance to be processed.

Figure 3-6 showed a P time for American automobiles of 11 weeks and the comparable P time for Toyota of 3 days. If Q is typically 95 percent of P time, it is easy to calculate that removing it would allow American automobile manufacturers to have a P time of 11 weeks $\times$ 5 days/week $\times$ .05 or 2.75 days. Remarkably close to Toyota at 3 days, don't you think? Toyota has a better "oil refinery": their products are continually flowing, not resting in a queue somewhere.

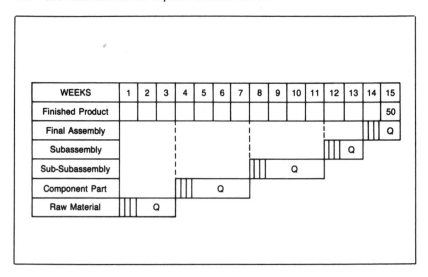

WEEKS	1	2	3	4	5	6	7	8	9	10	11	12	13	14	15
Finished Product															50
Final Assembly															Q
Subassembly													Q		
Sub-Subassembly									Q						
Component Part						Q									
Raw Material			Q												

FIGURE 8-4. Queue's Impact on P Time

Three Reasons for Long Queues—
And What You Can Do About Them

There are three primary reasons for long queues. Each must be understood and attacked aggressively.

1. *Inadequate capacity.* It is obvious that a resource is guaranteed a queue if it can produce (flow product) at a rate X, but work arrives or is scheduled at X plus. If this is a continual condition then the queue will grow and grow.

Here's a clear reason for my earlier admonition in chapter 7 about not stuffing the master schedule. Overpromising the plants' capability can only result in long queues, long P times, and disastrous business results. It's the equivalent of gridlock. And now the crisis expediting process, akin to emergency vehicles forcing their way through the traffic, will create even more disruption and, believe it or not, less total output for the business.

This is true whether the resource is a factory work center or a vendor. We can manage the capacities of our own work centers, but we often don't know anything about our vendors' capacities and we certainly don't manage them. It is obvious that a deeper involvement in our vendors' affairs is necessary if we want to avoid the inadequate capacity problem throughout our logistics chain.

2. *Erratic flow*. A resource can have enough or even too much capacity on the average but the work demands are erratic, so the capacity is overloaded at certain times and under utilized at others. Think of a highway during rush hour and again at 3:00 A.M., for example.

Figure 8-5 shows the actual receipt of work into a work center that has a capacity of 270 hours of work per week. The average receipt matches the capacity almost exactly, so the master schedule is about right on the average. But look at the peaks and valleys!

This has been dubbed "the pig through the python" effect. A python that swallows a pig has a large visible bulge that slowly moves down the python's body. A large batch of work that exceeds the average capacity of the resources it uses results in a large bulge in work-in-process that slowly moves through the factory.

It is obvious that a factory runs on the average. About the same number of people come to work each day and work about the same number of hours at about the same efficiency and with an average Murphy factor to contend with. Hence, their output is about the same on the average.

Week Number	Hours Received
1	286
2	50
3	147
4	176
5	695
6	531
7	139
8	321
9	61
10	284
Total	2690
Average	269.0

FIGURE 8-5 Erratic Flow Example

It's also true that manufacturing supervisors are measured on their use of resources. Idle time is a heinous crime, but overtime is acceptable. Faced with this bias, it's obvious what a supervisor must do under erratic flow conditions—carry a large backlog (queue) at all times to smooth out the peaks and valleys of demand. If the supervisor of the work center in Figure 8-5 doesn't have over 400 hours in queue at the start of week 1 (about 1½ weeks work at current capacity) there will be idle time in week 4. He'll prevent that at all costs.

By now it is obvious that actual lead times are erratic for products going through this work center, depending on the queue and the erratic flow. Imagine products going through five work centers like this one and you can understand the unpredictability of the process. Actual lead times will be all over the lot, sometimes short, other times long. Why this condition exists may have been confusing to you in the past, but now it should be crystal clear.

Force a smooth flow by balancing the master schedule. It is obvious that if a work center runs on the average, then the flow of work demanded from that resource should also be average. The alternative is to make work centers capable of adjusting their flow rates quickly to suit the demand.

There are two fundamental scheduling reasons why work centers receive work as in Figure 8-5. First is the way batching decisions are made, thanks to our old friend, the EOQ. A large batch created because of a long changeover is analogous to the pig through the python. Reducing changeovers therefore smooths out the flow. What did I tell you about the synergy of quick changeovers?

Second, an unbalanced master schedule will result in unbalanced (erratic flow) upstream work centers. We talked earlier about a feasible, doable, realistic, master schedule but not about its balance. To demonstrate balance, suppose this month's task is to make 20,000 product As, 10,000 product Bs, and 10,000 product Cs. It is a four-week month. We have the capacity to produce 10,000 of any of these different products per week, so this plan is realistic.

The normal way of scheduling these products would be to determine which has number one priority, which is second and third, and schedule them accordingly. Assume A is more critical than B and B more critical than C, either because that's the way customers want them or that's the priority needed to replenish inventories. In that case, we would schedule 10,000 As in each of weeks 1 and 2, 10,000 Bs in week 3, and 10,000 Cs in week 4, as shown in Figure 8-6. But this will cause unbalanced demands on feeding resources because these are different products with different components processed differently on different resources. Although the master schedule is realistic, it is not balanced.

An improvement is to make some of each product every week. This means dividing the batch sizes by four. With long changeovers this is not feasible, because of the lost output this would cause. But if the changeovers have been reduced, this will be no problem.

Producing some each day is obviously better than each week, some each hour rather than each day better still. Utopia is one unit at a time, set into a cycle of repetition that never varies and extends out the full month. But you need to achieve NTED to get there. If this was your master schedule and no batching was done at intermediate production levels, then it is obvious the demands on all feeding work centers would be perfectly level. Although few companies will achieve this utopian condition, I am sure the value of reducing changeovers by any amount to improve balanced demands is clear.

The amount of erratic demand that is left must now be eliminated by making the producing resources flexible. If the master schedule is realistic, and it obviously must be if you want excellence in a manufacturing concern, then you have enough resources on the average. The only question is, "How flexible are they to the various demands placed on them?" Here is where cross-training programs pay off. It's also where some our traditional objectives—for example, high machine utilization—are devastating.

Normal	Week 1	Week 2	Week 3	Week 4
	10,000 A	10,000 A	10,000 B	10,000 C
Better	5,000 A 2,500 B 2,500 C	5,000 A 2,500 B 2,500 C	5,000 A 2,500 B 2,500 C	5,000 A 2,500 B 2,500 C
Getting Close	Day 1	Day 2	Day 3	Etc.
	1,000 A 500 B 500 C	1,000 A 500 B 500 C	1,000 A 500 B 500 C	1,000 A 500 B 500 C
Utopia!	ABACABACABAC . . .			

FIGURE 8-6. Scheduling Alternatives

3. Tinkering. This is the slang expression used to describe the lead-time syndrome, a very real and insidious phenomenon within logistics systems. In simple terms

lead time = order backlog ÷ flow rate capability

We put the respective value of lead time for each item into our logistics planning systems to trigger the release of replenishment orders.

But what happens if the order backlog increases slightly when flow rate capability does not? The answer: actual lead times grow. Because the system was using the previous, shorter lead time as its trigger method to release work, this means jobs finish late.

Some people immediately see this as a planned lead-time problem. It is obviously too short. But if you lengthen the planned lead time, because logistics systems always schedule backward, you simply release more work sooner. This adds to the order backlog, making actual lead times even longer. Queues of work grow, expediters "magically" appear to "solve the problem," and the problem gets steadily worse.

This syndrome is most visible in purchased materials, although it is also applicable to manufactured parts. Lead times for many commodities regularly fluctuate way out of proportion to the variation in demand for them. Bearing lead times, for example, go from 4 weeks to 40 weeks and back again. Once they even went to 104 weeks and back! Semiconductors have been on this roller coaster since their invention. Castings, steel, aluminum, and many other commodities regularly exhibit these traits.

In fact, the companies that make these commodities are extending their D time, often far in excess of their actual P time. I remember getting a quoted lead time of 45 weeks for resistors when their P time was about 4 weeks.

The insidious part of this is how customers now react. In chapter 5 I talked about the four characteristics of forecasts. One of these was that the further out you go, the more wrong you are. Figure 8-7 shows this statement conceptually. All orders, whether purchase orders, shop orders, schedules, or customer orders, are generated by one of three wrong forecasts. You or your customer forecasts an item needed (wrong!), the quantity needed (wrong again!), and the date required (probably the most wrong of the three).

As an inventory clerk or buyer responsible for a product with lengthening lead times, you now have two choices:

1. Play it straight. Order just what you think you need and hope to react quickly to make corrections when you are wrong.

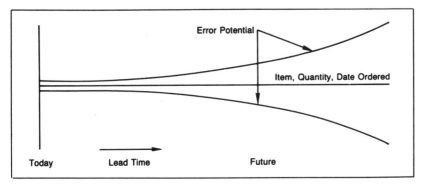

FIGURE 8-7. Triple Forecast Error Problem

2. Hedge your bets by buying more items, in larger quantities, for earlier dates. Yes, this will result in higher inventories, but you are normally penalized far more for shortages than excesses, so you are covering yourself better by hedging. And the orders won't create inventory unless they are allowed to mature into products in the factory. So the game starts of placing orders in excess of need and then rescheduling or cancelling them as the real needs become clearer.

The futility of this is apparent when you consider what is happening to the order backlog. More orders for more things earlier than needed means the backlog grows, forcing out lead times and creating more need to cushion more which means the backlog grows, and so forth. When vendors increase their lead times, the impact on your P time should be obvious.

APPLY FACTORY LEAD-TIME SOLUTIONS TO THE TOTAL LOGISTICS CHAIN

I have used some specific examples to demonstrate the reason for long P times and in some cases suggest solutions. I will discuss additional solutions in the next chapter. But first I want to make sure you see how these ideas apply outside the narrow examples I have described.

How Your Selection of Vendors Affects Your Lead Time

The lead-time syndrome is very real where vendors are concerned. Vendors and customers communicate with purchase orders and sales orders for specifics, ordered as they relate to lead times. But the problems of inadequate capacity and erratic input also apply to vendors.

If a vendor has more business than can be produced in the quoted lead time, or if orders arrive from customers erratically (as they usually do), then the vendor has no choice but to have long lead times.

Disjointed location and selection of vendors is analogous to poor layout in a factory. Buying from many sources scattered around the world in a search for material cost savings is bound to increase P times and increase the risk that some items will not arrive on time, even with safety or hedge inventory.

One of my clients in Vienna, Austria has launched an aggressive program to get all purchased parts sourced within 300 miles of the factory. At the moment, many items are procured in the Far East. But the plant manager now feels that the costs resulting from long lead times, tougher communications, more inventory, and higher transportation outweigh the wage rate savings.

Manage Engineering to Reduce Lead Time

Many products need some custom design engineering before procurement and/or production can start. Examples are oil well production devices, specialty motors, ships, custom-built machine tools, metal buildings, and telephone exchanges. This puts engineering on the critical path and by definition makes it part of the P time. Engineering designs are often late, and it takes too long to get specifications for long lead time items.

The major reason for this is not technical. It is simply that this resource, and associated ones like drafting, have rarely been thought of or managed in the same vein as a production resource. But engineering is subject to the same problems of inadequate capacity and erratic input as a factory or vendor. Their only choice under these conditions is a long queue and a long P time with an erratic finishing schedule.

Treat design engineering as a production resource. Plan the capacity and flow work into design engineering at a smooth rate equal to that capacity. You will be surprised at the reduction of engineering lead time you get.

At the same time, consider scheduling engineering to speed up their throughput. Most engineering departments break up their projects or customer orders between engineers. One engineer works on one project or order.

To demonstrate that there is a better way, let's look at an analogous situation: unloading ships at docks. Six ships arrive at the same time to be unloaded. We have six wharfs for them to tie up to and six cranes to unload them. One crane can unload one ship in six days. We can assign multiple cranes to a ship and reduce the unloading time proportionally. How should you assign cranes to the ships to minimize the turnaround time?

Figure 8-8 shows the two extremes, one crane per ship or six cranes per ship. The six-cranes-per-ship method results in faster turnaround for every ship except number six. What does this say about allocating engineers to projects or orders? Apply as many engineers to one project as possible. Back away from this idea only if the engineers get in each others way. When one project is complete, move onto the next.

Beware the political problems this decision may create, however. You won't be able to tell all the key persons interested in your engineering projects, "Yes, someone is working on that." You will be able to say that only for whatever project is currently getting attention. But the total throughput time for all projects will be dramatically improved, and all customers will get their orders or projects designed sooner as a result.

Use this same approach with all projects —for example, systems designs—whenever resource interference doesn't occur. Assign everyone to one task rather than one person to a task. You'll handle demands much sooner that way.

Treat All Resources Like Production Resources

Tool designers, tool makers, industrial engineers, salespeople, and so forth, are all resources with a certain capacity on which certain demands are placed. They too face the problems we've discussed. If

Ship Number	One Crane per Ship-days in Dock	Six Cranes per Ship-days in Dock
1	6	1
2	6	2
3	6	3
4	6	4
5	6	5
6	6	6
Total	36	21

FIGURE 8-8. Two Ways to Unload Ships

they are part of the critical path for your products, then they will extend P times for sure.

One of my clients in Europe achieved significant P time reductions after focusing on the sales order specification process and treating it as a production resource. This was their largest bottleneck and there was no other way around it. The company implemented some rudimentary but effective measures and controls and expanded capacity in this area. As a result, on-time delivery performance to the customer jumped dramatically while delivery lead times were shortened drastically.

Cutting P Time by Streamlining Distribution

We have shown how JIT concepts work in a factory and vendor setting. It is not so easy to see how they apply to distribution. One particularly relevant application—location—has already been considered, in terms of both the proximity of the plant to the market and to vendors. This same concept of proximity also applies to the distribution network.

Using air freight rather than truck, barge, or sea freight can compensate to some degree for distance, but usually at a cost penalty. If the cost penalty exceeds the inventory, packaging, and handling savings, air freight is tough to justify.

Erratic flows and large batches. If you look at the distribution network in more detail you often find the same jumble of erratic flows and crossing paths that you find in a functionally laid-out factory. This arrangement is usually designed to achieve freight savings by sending full truck loads instead of less than full truck loads.

Like stockrooms in a factory, warehouses, too, are traffic jams on the highway flowing product from vendor to customer. In consumer products, the average distribution time often exceeds the total P time of procurement and production.

Figure 3-2 shows this buildup of total P time from vendor to customer. The proportion consumed by distribution is 33 percent, a worthwhile target to reduce.

Inadequate capacity of transport, erratic demands, and large batches to get freight savings are some of the reasons for this extended distribution time. We'll discuss the specific solutions in the next chapter.

9 | How to Achieve Manufacturing's Top Three Goals: Increase Output, Reduce Costs, and Generate Cash

P times for utopia are easy to describe. They should be zero. Products should be produced instantaneously out of raw materials that can be procured instantaneously. Anything that moves a manufacturer toward utopia is positive; anything away from it, negative.

Eli Goldratt defined the objectives of manufacturing in his book, *The Goal*, (North River Press, 1984) as:

- Increase throughput (total output of product).
- Reduce costs (all).
- Generate cash.

It is obvious that reducing P times will generate cash by decreasing inventories. It is not quite so obvious that reducing P times will also reduce costs and increase throughput.

But if you reflect on some of the problems that cause long P times, such as inadequate capacity and erratic demands on resources, it's obvious that solving these problems will increase throughput. And reducing changeover times, taking away the costs of expediting to handle the latest crisis, and simplifying scheduling by focusing on cells rather than functional groups, will reduce costs. I am sure you can envision other ways in which reducing P times will help achieve these three objectives. I'll point more out later in this chapter.

HOW TO ENSURE SMOOTH FLOWS FROM RAW MATERIALS TO MARKETPLACE

I used the oil refinery example in chapter 2 to suggest a utopian condition for a plant. Materials flow smoothly, reliably, and quickly into the plant, are transformed and shipped immediately to customers. Throughput (P) times are short and flexibility is high. We are now managing a controlled process.

But the marketplace often is, or at least appears to be, unstable. The mix of products ordered changes and so does the volume.

It's obvious that with adequate flow rate capability and short P times, mix changes can quickly and easily be handled. It's also true that mix changes are the most common and volatile aspect of the marketplace. Through mix flexibility, you can solve your most common problem.

THREE WAYS TO HANDLE VOLUME FLUCTUATIONS

A different approach is required for volume changes. Clearly understand the pros and cons of each choice before implementing the best decision for your business.

Solution 1: Build Finished Goods Inventory

Use this inventory to decouple the smooth production of the factory from the volatility of the marketplace. The inventory acts as a buffer and absorbs the variability in demand.

Of course this works only if the sales volume frequently varies both above and below the factory's smooth output. If variations are infrequent or the variability in demand is biased toward one side of the factory's production, then finished goods inventory will not solve the problem. The amount you need to provide the cushioning will be excessive.

Keep in mind too that inventory always consists of specific items. For finished goods inventory to be able to cushion demand swings, it has to be the *right* inventory. As we know, forecasting the right things to have in inventory is not an easy task. Using finished goods inventory as a decoupling mechanism often brings its own set of problems—high inventories, wrong inventories—and it works only under some specific, limiting conditions.

Solution 2: Synchronize Production Rates with Demand Rates

In other words, be able to quickly expand or contract the pipes in the oil refinery. This solution has implications for people, mainly direct labor, and machinery. Let's consider direct labor first.

Just how costly is direct labor? The cost of direct labor is now averaging only about 7 cents of the sales dollar. The value of keeping this cost low, even though still somewhat important, is not as important as it was back in the 1930s when the cost was closer to 30 cents. Some companies are even beginning to consider direct labor as a fixed cost. They attack the other costs of building inventory, long P times, and NVAW, which are far in excess of the labor costs.

And isn't finished goods inventory, our first choice above, in fact idle direct labor? Whenever you see boxes or racks of inventory, you should visualize piles of raw material next to some direct labor workers with their arms folded, staring back at you.

In the case of inventory, though, you spent cash earlier than necessary for both the labor *and* the raw materials (the latter close to 40 cents of the sales dollar) and combined them into a specific item. This is directly opposed to one of manufacturing's objectives: to generate cash. You now spend money for warehouses, for people to store and handle the inventory, and for systems to manage its replenishment. This is also in direct opposition to another manufacturing objective: reduce costs. And if you made the wrong things, which will show up in your obsolescence costs and discounting of slow-moving inventory, you didn't increase throughput either.

Would it have been better to spend slightly more on direct labor to provide the flow rate flexibility to meet the marketplace? In many cases the answer is "yes." Overtime, temporary hires, and a cross-trained, flexible work force are ways of minimizing the costs of flexible labor. But be prepared for additional costs in this area to get the refinery in synchronization with the market and achieve manufacturing's real objectives.

Warning: high machine utilization could be detrimental to your profit picture. Machinery decisions are also crucial. If you want to be able to vary output rates, then some excess, and at times idle machinery, is mandatory. This flies in the face of many people's views that high machine utilization is good. But high machine utilization means little flexibility for changing production rates.

Again, your choices are to carry inventory or have some excess idle machinery. But is this really a choice? Isn't inventory in fact excess idle machinery in a different form, just as labor was? Inventory certainly is an idle asset. And which way are you better off? Should you keep the machine assets low to have high utilization but now need inventory to satisfy changes in market demand? Or would you be better off putting the inventory value into excess, flexible machines that run only when needed? In most cases your business results would be better with extra machines and less inventory. At least you can depreciate the machines.

Solution 3: Cultivate a Group of Subcontract Vendors

Another way of solving volume changes is to have some subcontract vendors qualified to make your products who can quickly adjust their output rates to absorb your peak demands. Some companies find it worthwhile to pay vendors a retainer for capacity, whether they use it or not, to get this flexible reaction to change.

FOUR KINDS OF FLOW RATE

I have used the term "flow rate" in preference to the more traditional term "capacity" throughout the book. But a further definition of these words is now necessary. The first thing is to separate two other terms that many people confuse: workload and flow rate (capacity). *Workload* is the amount of work waiting to be processed by a resource. It is the same as queue or work-in-process. *Flow rate* is the amount of work coming into or leaving the resource.

If you agree we want short P times, then workload must be close to zero at all times. Product flowing into the resource must immediately flow out. But there are four kinds of flow rates for every resource and it is critical we understand them all. The resource can be a machine, group of people, vendor, or factory in total.

1. *Theoretical.* This is the flow rate a process could produce under ideal conditions. It is what the salesperson who sold us the machine or process said it should be able to produce but never has. This information is useless for planning and scheduling a plant or promising deliveries to customers. Murphy will make sure you never achieve this rate.

2. *Required.* This is just the right amount of good product flowing to exactly match what is needed to execute your master schedule. Flow more than this and you create inventories, less than this and you cause shortages or backorders.

3. *Demonstrated.* This flow rate is based on recent historical average output of good product. If you make no new management decisions, this is your best guide to how much will be produced in the future.

4. *Effective.* That portion of the demonstrated flow that is the right amount of the right things flowing quickly and directly to the customer. Hence the balance, ineffective by definition, flows into inventory. You hope to use it sometime in the future.

It should be obvious that utopia is when today's demonstrated equals today's required and that it is all effective. Any imbalance is going to detract from manufacturing's objectives.

One result of an imbalance is obsolete or slow moving inventory. This means that demonstrated capacity had to be higher than that required to support customers, the difference thrown away or sold for recovery value. Costs are obviously higher, cash is being consumed instead of generated, and throughput is not enhanced. Throughput could even be reduced, because you are using scarce productive resources to make things that are later thrown away.

Several things make demonstrated capacity ineffective. Our old friend, lot sizing, always causes this because we make in batches, even though the demand is so many per day or week. Look at the General Motors numbers from Figure 8-2 and consider the ineffectiveness of building 20 days' supply of parts. Several years ago GM's Australian affiliate, Holden, because of their lower production volumes, produced batches equal to an average of 45 days' supply! How ineffective can you get?

Another reason why capacity is ineffective is long P times—forecast error guarantees ineffective production. In fact, inventory of any form guarantees ineffective capacity. Mather's first law of inventory is, "The day you put something into stock it's the wrong item." There's little or no value in that, even though you made it "efficiently."

EFFECTIVE VERSUS EFFICIENT FLOW: THE CRITICAL DIFFERENCE

It's important that we separate two ideas that many people, especially senior managers, confuse. I discussed these briefly in chapter 1. These are efficient use of resources and effective use of resources.

Efficiency is a term defined largely by the financial reporting system. It is calculated as:

Efficiency of a resource = output ÷ input

There is no detraction from this measure if the resource makes more or less than is needed or the wrong things. The blind assumption is that the production is correct.

But let's get really specific about this measure with a simple example. Let's say we have three machines in a line, processing products in sequence.

We have calculated that, at rated speed, the first and third machines need eight hours to produce what's required for the master schedule, the second machine only four hours. The reason is that the first and third operations take two minutes per piece, the second only one minute. I am the supervisor of the area. Because of the varying demands on the three machines, I have decided to run the second machine at half its rated speed or to cycle on and off half the time to stay in balance with the two machines on either side. This means I must keep a direct labor worker at the machine for four more hours than is necessary, even though I have other work this operator could do. To do this other work I will bring in and pay a temporary worker for four hours.

The reaction from most people when I say this is, "You're crazy! That's not efficient. You have to pay for an extra four hours of labor. Your costs will go up and make you noncompetitive." There's no question these comments are valid, up to a point. But what's the other side of the coin?

Why High Efficiency Can Cause High Costs

If we run these machines out of balance at their rated speeds, the first machine must create a minimum of four hours worth of work-in-process inventory before the second machine starts up. This takes space, storage containers, and racks, which all must be bought. It also requires material handling, either from an indirect worker, or allowed for in the direct labor worker's production standard.

The second and third machines can start up together. But as the second machine is twice as fast as the third, it has processed the full day's requirements in four hours. The third machine has another four hours to go, so again, four hours of work-in-process inventory has been created. More space, racks, storage bins, and material handling.

Quality is a concern. And now enter another consideration, quality. Let's assume that the first machine can barely make products to specification, either because the tooling is worn, materials are off spec, or the operator had a big night last night. Because of this, some percentage of bad parts get made. These are now in the pile of work waiting for machine two to process. Machine two starts up and the operator has no problems for a while. But then she picks up a bad part for processing. Let's assume the operator immediately realizes it is bad. Now what to do?

It's obvious she won't process it. What may not be so obvious is that because she is measured on efficiency, she will set the bad part aside and reach back into the pile for another piece. As most of them are good, she will continue working, setting aside bad pieces now and again when they are found. At the end of the shift or batch, these bad parts will be routed to quality control for disposition—which means more material handling, paperwork, and space, but most importantly, more delay.

At the same time, let's assume some bad parts slipped past this second operator and were processed further. They are now in the pile ahead of the third operator. The second operator also made a few errors during her operation. So the third operator will play out the same scenario. The final inspector, the customer, is bound to receive some portion of these defective products.

Quality control's job is to find the cause of the problems and fix them. But a few hours or days will have elapsed between the problem's occurrence and the quality control analysis. By then, the trail will be cold. Hence the probable outcome is that the bad parts will be scrapped or reworked, and that's that. But bad quality is inefficiency of the worst kind, because it not only wastes labor, it also wastes materials. And reworking wastes even more labor.

So what are the benefits or costs of high efficiency? Yes, we do save four hours of direct labor. But we incur the costs of extra space, storage containers, racks, inventories, material handling, and poor quality.

The throughput (P) time is also affected. On average it will be eight hours, whereas the actual processing time is only five minutes. This is not the way to get the P:D ratio less than 1 to 1.

How to Lower Total Costs Through Inefficiency

Let's see what happens if the machines are perfectly balanced, and yes, for now, assume the second machine needs the operator for four more hours to achieve this. All machines start up at the same time. There is only one part in process between each machine, why do you need more? Therefore the machines are close together, saving lots of space. No racks, storage containers, or material handling is needed.

Let's say the first operator makes a bad part, for whatever reason. The second operator immediately picks it up and realizes it's bad. She obviously must stop work. It won't be long before the third operator also stops work as there is now no flow of parts to this machine or any pile of inventory to choose from.

Now we have idle time—as you know, a heinous crime in manufacturing. There is only one thing worse—bad quality. So now the operators get together to solve the problem that caused the bad part. If necessary, they call in technical support. The trail is hot so they know the conditions of material, tooling, machinery, operator skill, and product specifications. Hence the chance of identifying and fixing the problem is high, preventing its recurrence. And the real efficiency of all three resources increases steadily as product quality gets better and better.

Speed the flow to improve the quality. Quality is a key objective for all manufacturers and an increasingly important one in the face of excellent foreign competition. Many companies have instigated quality improvement programs to attack this very real concern. But improving quality has no direct influence on inventory levels or throughput times. As such, it is a very narrow objective.

Reducing inventory levels and speeding up the flow not only generates cash, it also improves quality. In fact, the best quality improvement program you could start would be to force inventories down. Quality has to improve or the factory will stop.

GO FOR A BALANCED FLOW

Figure 9-1 summarizes the benefits of unbalanced versus balanced flow in this simple example. Consider the two P times, one with an average of eight hours with unbalanced "efficient" flow; the other only five minutes with a balanced flow. And the effective flow will be higher with a balanced flow.

It should be obvious that, in many cases, excess labor is preferable to excess inventories. For example, if you could reduce work-in-process inventory by $1 million with better balance, how many direct labor workers could you afford to have idle all year long? The answer depends on how you value the costs of inventory. Are inventory costs simply financing costs, or do you view them as lost opportunity costs?

Unbalanced	Balanced
Less direct labor	Improved quality Short P times Less inventory More flexibility Reduced expediting Less material handling Less space

FIGURE 9-1. Benefits of Unbalanced Versus Balanced Flow

Even at a conservative value of 10 percent per year, this is a $100,000 cost. Dividing this by your out-of-pocket costs for direct labor wages and benefits will give you the equivalent number of people. Of course, using temporaries or supplying the equivalent amount of extra capacity through overtime would make the numbers even more favorable.

HOW TO ACHIEVE BALANCE AND COMPENSATE FOR EXCESS DIRECT LABOR

What if we could reduce the negative impact of excess direct labor at the same time as we achieved balance? Let's look at an example. I visited a company in California that worked three shifts in the production areas but only two shifts on the packaging lines. When I arrived at the packaging area, about mid-morning, I could see rack after rack of product waiting to be packed. This was the material made on the third shift while the packaging area was shut down. My guide on the plant tour was the head of industrial engineering.

My immediate question to him was, "Why don't you slow down the packaging line to two-thirds of its rated speed and run it three shifts?" The industrial engineer looked at me as if I had just crawled out of the woodwork. He said, "Why? We need only two shifts to package what the factory produces in three. Slowing the packaging line down would hardly be efficient."

I asked, "What crewing do you have in packaging?" He replied, "Five people each on two shifts." I said, Two-thirds of five people would be three and one-third people, possibly requiring four. (The operation was largely people paced.) Four people on three shifts would be twelve. But what if we could run the operation crewed with three people each on two shifts, four only on one shift? Yes, there would be a buildup of work waiting while the three people crews were working, but this would be significantly less than when the one shift was closed down."

My next thought was, "And if there is idle time while crewed at five people or material handling needed because of the large inventory, it may be possible to run a three-shift operation staffed with three people per shift and still package to suit production. Now that would really improve productivity."

A couple of weeks later I was told that the packaging line had been slowed down, it was crewed at three people per shift and was able to maintain output. The industrial engineer found plenty of wasted effort moving inventory from the large storage area to the packaging line. He also found that workers spent a lot of time searching for the most urgent thing to pack (usually buried behind later production), also causing wasted effort. The space saved will allow the factory to expand considerably as volumes grow without adding any additional buildings. In fact, I recently learned they are going to consolidate all production into one of two buildings and lease out the unused building. A good part of this revenue will be due to balancing flows and not being so concerned about efficiency.

In this case, there were no direct inventory savings. The items were simply packed and moved into the finished goods warehouse. But the warehouse also now works three shifts, staging the factory's production into shipping batches. As soon as trucks arrive in the morning they are quickly loaded with the right products and sent on their way. This has improved truck turnaround time, in many cases has lopped a day off the time before products are available, and has improved customer service considerably.

Four Choices for Using Excess Direct Labor

Returning to my previous example of the three machines where I discussed balance or efficiency, what can we do with the four hours of excess labor on the second machine? There are four choices that come to mind.

Choice 1: accept this direct labor cost as a fixed cost. Get your benefits from less inventory, a shorter P time, and better quality.

Choice 2: fill the second operator's time by having her do other jobs. These could include preventive maintenance on all three machines, material handling in the area, or statistical process control. Maybe, by relocating another machine close by that also needs to run only four hours per day, she can alternate between the two machines. Now she is effective and efficient.

Choice 3: re-engineer the first and third operations to also perform the second operation. Eliminate the second worker altogether. Warning: Make sure that line balancing doesn't hide idle time. This throws a different light on "line balancing," that process where work is evenly distributed over all workers on an assembly line or in a machining cell so that they all execute a given rate of output. For example, line balancing would have attempted to give each of our three workers in the previous example 1⅔ minutes of work per operation instead of the 2 minutes, 1 minute, 2 minutes that we started with.

But this would have hidden the four hours idle time of the second operator as one and one-third hours for each. It's unlikely, therefore, that the idea of removing four hours of labor to eliminate the second operator would have been so apparent.

This doesn't mean I'm against line balancing as a concept—far from it. But it is impossible to have a line perfectly balanced—imbalances will always be present, caused by different skills and learning rates of the operators, slight changes in needed output rates, process inconsistency, and so forth.

What I am saying is that any excess time on a line should be kept on one operator. Have all the others operate at full capacity. In fact, deliberately unbalance the line so the lost motion is fully visible. Now work to improve all the processes to eliminate this partial job. Removing an operator will really increase productivity. Spreading the lost motion over several operators won't.

Choice 4: automate the second operation. This could be done with a robot or automatically operated machine or process. I find it amazing that as soon as I ask, "What if the second operation is automated?" everyone says it's OK to slow the machine down. Didn't automating the second machine cost money? How come it's acceptable to slow down an automated process but not a manual one?

Select the Best Choice for You. These four choices need to be weighed carefully and applied to your particular circumstance. Make sure, however, that no matter which choice you make, keep balanced flow rates uppermost in your mind. Don't let efficiency come first. Remember that the soft benefits of short P times and increased flexibility along with the hard benefits of less inventory and better quality, often far exceed any direct labor savings higher efficiency may bring.

BEWARE OF MACHINE UTILIZATION PRESSURES

High machine utilization, a measurement often assigned to the production people is another invalid concept. High machine utilization is assumed to be good, presumably because it amortizes the machinery faster.

George Plossl, a well-known figure in the field of manufacturing, once told me a story from the time when he was promoted to plant manager. The company had recently justified and purchased an expensive machine tool. The general manager had decreed that the machine should never be idle. Shortly afterward, George was presented with the justification for another one of these machines, largely because the existing one was overloaded. He was getting ready to add his signature to the machine appropriations document when a thought struck him. What work is currently on the machine and do we have to produce it all this way? Could other machines that were not busy perform some of the work?

A quick investigation showed that the industrial engineers and production supervisors had taken heed of the general manager's decree and routed parts across this machine "to keep it busy." But many of these parts could be made cheaper and better on the traditional machine tools that were now operating way below capacity.

The parts were rerouted to their most efficient process and this "overloaded" machine suddenly had idle time. But now the financial people got into the act. They cried that costs would go up, because with less operating hours, the amortization cost of the machine per hour would increase. They were the prime movers behind the general manager's edict for keeping that machine busy.

Don't Confuse Sunk with Variable Costs

There are sunk costs and there are variable costs. If you have a machine idle, it is because you bought too fast a machine, or because you want the production expansion capability, or because it's the only machine available to perform a certain operation. The only out-of-pocket costs you incur by running a machine you own or leaving it idle are the costs of raw materials or direct labor while it is running. Hence, you should try to keep all machines idle. Run them only just enough to support the orders you get.

By forcing the machine to run above the flow rate required to execute the master schedule, just to amortize its cost and increase utilization, you are simply transferring a machine asset into an inventory asset. How can transferring an idle machine asset into an idle inventory asset benefit the business? The answer is, of course, that it can't. It hurts the business by building something customers may not want, consuming additional company cash for the raw materials, creating a more complex business environment to manage, and adding lots of NVAW.

The department which *should* be measured on machine utilization, if any, is sales. Sales determines which machines run and at what rate through the volume and mix of orders they book. The best the production people can do is make sure the machines run exactly at the rate to support sales.

Approach Machine Utilization Differently

Machine utilization is not a completely erroneous objective. But it must be approached from a different angle. Earlier I talked about the four kinds of flow rates: theoretical, required, demonstrated, and effective. I stated that required today should be demonstrated today and it should all be effective.

But how should these relate to theoretical? Well, it's obvious they must be below, by some margin. How much depends on three factors: what you need to execute the master schedule, what nonoptimum conditions you will experience, and how much flow rate contingency you need.

Required capacity is the rate that must flow to execute the master schedule. This number, divided by theoretical, is the machine utilization. It's not a good or bad number, it's the facts. If we can't smooth out the flow, then some flexibility contingency must be provided to take care of the peaks. This, by definition, says that idle machine time will be a fact of life and a necessary one.

Nonoptimum conditions and necessary idle time will always detract from theoretical capacity, leaving practical capacity. Quality problems, machine breakdowns, idle time incurred while operators wait for materials, changeovers, and so forth all reduce theoretical capacity.

The gap between practical and required plus contingency means the machine will be truly idle and hence a nonbottleneck process. There is no value in trying to use this capacity by running the machine more hours. It's the master schedule that needs attention, not this machine.

Eliminating the nonoptimum conditions or necessary idle time is also of no value for this resource except for the direct costs it reduces or direct benefits it gives. All it does is to increase the amount of spare capacity, making it more of a nonbottleneck.

For a bottleneck resource, though, attacking the nonoptimum or "necessary" idle time will really pay off. It allows the required capacity to expand toward the theoretical and still be feasible. This will increase machine utilization to be sure, but in a good cause, not as an objective in itself.

Be careful of machine utilization as a production performance measure. Understand that low machine utilization usually comes about because your sales mix or volume has changed or because you bought the wrong machine. You won't fix either of these problems by forcing the utilization higher, regardless of what the financial system says. If you follow their logic you will move further away from manufacturing's objectives, and that doesn't make any sense.

An effective resource is one that is making exactly the right amount of exactly the right things at exactly the right time, all of good quality, to execute the master schedule. Don't let anyone tell you otherwise. They are confusing effective use of assets with efficiency.

GUIDELINES FOR PLANNING FLOW RATES

Capacity is an imprecise term. The best you can do is plan on averages unless you have a very controlled process and you can remove the human variables of efficiency, absenteeism, idle time, and so forth.

Many people don't agree with this statement, however, and feel that capacity can be precisely calculated. Industrial engineers with stopwatches and time standards to five decimal places are probably to blame for this view.

Many supervisors are looking for the exact demands on their resources several weeks or even months into the future. No one can argue with this objective, except that it is unachievable. How many master schedule, engineering, and routing changes will be made between now and then to prevent today's prophecy from coming true? The detailed calculation will be expressed in terms of standard hours required based on today's business conditions because that is the level of detail the system contains. Comparing these standard hours based on today to the average expected capacity that will be demonstrated sometime in the future, considering all the variable problems mentioned earlier, will be futile. It's a search for precision, but in the search you will lose accuracy.

What we need is a useful unit or units of measurement that can convert your master schedule into required flow rates at all critical upstream resources. As described in chapter 4, we have to be able to size the oil refinery's pipes. This unit or units of measurement must in fact be measurable. We have to put flow meters in the pipes and determine if they are actually flowing product at the required rate. Our flow meters must quickly sense deviations so corrective action can be taken.

Where capacity is concerned, you must always have a little more than you need. Never have less. My comments about short P times and not stuffing the master schedule should have persuaded you of this by now. If not, please go back and reread these sections so that you understand why investing in a little more machinery can mean investing in a lot less inventory. The net result will be a big gain in asset turnover.

Close Enough Is Good Enough: Rough-cut Capacity Planning

If capacity is not a precise subject, it is obvious that approximations are close enough. This is why the following techniques are generally called "rough-cut capacity planning."

We start with the master schedule. We could convert this into financial terms: net sales value, cost of sales value, or some deflated sales value. Maybe we could convert it into units by broad family; into equivalent units using some statistical weighting factor; or into tons, feet, or a unique industry-accepted measure. This will define in flow rate terms our required rate of product movement out of the plant.

We now compare these figures to recent historical average actual output (demonstrated) using the same unit of measure. If they match, chances are the upstream resources can support this master schedule. If they don't, you have a problem that needs management attention. Decisions must be made to change the demonstrated or the required flow rates. If you choose to change the required flow rates, of course, you must change the master schedule.

This use of past history as the starting point for capacity analysis bothers many people. They ask, "What if there was a recent bad problem, for example, machine breakdowns? We shouldn't include this effect in the numbers, should we?" Or, "We are way below budgeted efficiency numbers. We'd have lots of capacity if only these would improve. So we'll sit tight and hope the efficiency improves quickly."

What these comments and questions show is a fundamental lack of understanding of flow rates, what affects them, and what changes them.

Flow rates are determined by the number of people on the payroll, the hours they will work, the resources you give them to work with, how efficiently they use these resources, plus the amount of nonproductive actions they have to do. Some of these factors can be changed quickly, but others take concerted effort, and above all else, time.

You can change capacity rather quickly by changing staffing levels, hours worked through overtime or short time, and subcontracting. These are also management variables, to be exercised as needed.

Although they also affect flow rates, efficiency, machine breakdowns, and the other variables of scrap and absenteeism cannot be quickly solved through management decisions. To attack these problems takes time and a very directed effort to influence flow rates.

The master schedule is product we need to ship today, tomorrow, next week, or next month. It's no use putting your faith in long-term solutions if you have a short-term problem. Hence you have to start from demonstrated average output as your base. If short term required flow rates deviate from this average output, start exercising your management prerogatives quickly. Either make the necessary decisions that will change today's demonstrated to suit tomorrow's required or modify the master schedule so its required matches today's demonstrated.

A rough-cut capacity example where a useful measure worked. At one point in my career I worked for Gilbarco, a division of Exxon. Gilbarco makes service station equipment, primarily gasoline pumps. Their capacity unit of measurement is equivalents; in other words, a single pump (only one hose) is worth one, a dual pump (two hoses) is worth two. As far as machining, subassembly, final assembly, and test, this is a good unit of measure because a dual pump is really two singles in the same casing. But in the case of sheet metal, this was a lousy unit of measurement. You need half as much sheet metal to make two equivalents in a dual as you need to make two singles.

However, this measure never was a problem because the mix of single and dual pumps ordered by our customers stayed in a rather narrow range. If the mix had varied widely, we would have had to establish a different unit of measure for the sheet metal department.

Now for the beauty of this measure. Everyone understood it. Sales talked equivalents, accounting used equivalents, engineering related to equivalents, and production reported equivalents. Even the general manager understood equivalents! If we were currently building 140 equivalents per day, as measured, and a new plan was developed that required us to build 150 per day, everyone knew we had to add about 7 percent direct labor hours in all areas that made pumps. Overtime, hiring, or subcontracting to this level was immediately authorized. Hence, we were soon producing at the 150 per day rate.

This doesn't mean we were happy with our efficiency or indirect activities—far from it. But it does mean we clearly recognized the objective: to execute the master schedule. And we separated long-term actions from short-term actions to do this.

Take care when using advanced capacity techniques. As stated in chapter 4, there are more refined methods of calculating required flow rates to execute the master schedule than the simple ones described above. Resource profiles can develop flow rates by critical resource. Detailed calculations using routings can give capacity requirements by machine or skill group. Be wary of all these more complex approaches. Keep asking yourself, "Am I losing accuracy in my search for precision? Do these approaches help the entire organization understand the need for adequate flow rates balanced to the master schedule? Or am I confusing the issue by getting into too much detail?"

To fine-tune the next few weeks, these more detailed techniques are probably effective. But don't let them be used for the big picture. Use simpler and clearer methods instead.

HOW TO SMOOTH FLOW RATES

I talked about erratic flow in chapter 8. I defined the reasons as lot sizing (the pig through the python analogy) and an unbalanced master schedule. Of course, execution failures also cause erratic flow. Machine breakdowns, scrap, and late materials all cause peaks and valleys in the work of downstream work centers.

When it comes to planning smooth flow, though, where do you start? How do you get, or at least plan to get, a smooth flow through all resources?

The answer: you work backward from the factory/customer interface. If you can't make this smooth then you doom the plant to erratic demands forever. Now you are forced to carry finished goods inventory to decouple the nervous marketplace from the stable factory. Either that or the factory must carry excess capacity to execute to the variable demand rates.

The same problem occurs as you move upstream. If assembly demands are erratic on subassembly operations, subassembly stocks will be needed to decouple the processes. The same is true for fabrication and vendors. Stocks decouple each process causing long P times. The first thing to attack is the finished goods inventory. Two factors cause it: unbalanced demands from customers and unbalanced supply from the assembly area. Let's start with the unbalanced demands, we'll come back to unbalanced supply later.

HOW TO ATTACK UNBALANCED DEMANDS FROM CUSTOMERS

Uneven demands from customers will be costly. "Customers" in this context could be the end consumer, another factory that will further process your materials, or a distribution network. I'll tackle them one at a time.

Smoothing End Consumer Demand

Customers, whether for consumer products, industrial equipment, or business equipment, usually buy in batches of one or in very small quantities compared to the supply quantity. So why is this demand erratic?

Beware periodic sales targets. Many times it's our way of relating to customers that makes demand erratic. The computer company mentioned earlier has an order booking pattern that peaks up at the end of the month, at the end of the quarter, and even more at the end of the year. The normal pattern of sales in a quarter is 20 percent of the quarter's revenue in the first month, 30 percent the second, and 50 percent the third. The sales pattern during a month is 10 percent of the month's revenue in the first week, 15 percent the second, 25 percent the third, and 50 percent the fourth. They have to ship product no more than 48 hours after the order is booked, so the shipping pattern is the same as the sales pattern.

The major reason for this pattern is the measurement program for the salespeople. They are not paid commissions but they do have monthly, quarterly, and annual revenue targets. Toward the end of a measurement period, if they are below target, the salespeople push hard for additional orders and get them. In fact, this company pushes so hard to meet these revenue budgets that they have trained their customers to delay ordering until late in any period, especially late in a quarter. Customers get better discounts, longer credit terms, free software, or more equipment and services included in the sales price just so the salesperson can book the orders.

This company has successfully taken what must be a fairly smooth demand for a product—large computers—and made it erratic. You should see the almost finished computers in inventory early in the quarter to level-load the factory. The factory is approximately twice as big as it needs to be just to store these units. And you can imagine the scrambling and overtime it takes to finish them at the end of the quarter, which sometimes includes converting them from one version to another at a significant cost increase. How's that for lots of NVAW, caused by uneven sales patterns, causing even more NVAW in the form of inventories, overtime, space and converting expense.

If we want to make daily, we have to sell daily, and vice versa. If we expect factories to produce an even flow, we must also expect sales to sell an even flow. This means that revenue targets for sales have to be for smaller time increments, dropping from quarterly to monthly to weekly, and eventually even to daily.

Beware periodic billings. Here's another example of a company that is its own worst enemy when it comes to erratic demand. A large American division of a huge European corporation bills its customers only twice per month, usually on the tenth and twenty-fifth, a practice designed to reduce effort in accounting and data processing. Payment terms start from the billing date, not the product ship date.

Customers view this as a great way to get 15 days of additional credit. Hence, they buy and request the bulk of their shipments the day after the billing date. They buy only what is absolutely necessary as the billing date approaches.

What does this cost the company? Excess inventories, warehouse overtime, additional warehouse costs because only a very small percentage of orders can flow off the production lines direct to the customer, less cash flow, and so forth. This induced marketplace volatility could easily be removed if the company changed its billing procedures.

Evaluate all your sales activities. Everything in your sales relationships to customers must be reviewed. If your sales relationships are causing uneven demands, change them. Review Figure 2-10 to stimulate ideas where you could be going wrong. Don't amplify the inherent market variability. Don't let your discount structures, for example, cause the pig through the python problem. Give discounts based on annual buys, not on single large purchases. This review must encompass sales incentives, credit terms, and product offerings. This last item I'll cover in depth in chapter 10.

The changes you make in your market relationships must be done carefully. Customers are used to dealing with you in a certain way. Change this slowly, over time, to avoid rejection by the customer. But keep the changes coming with one objective in mind—a smooth continuous demand from the marketplace.

Note that nowhere am I attacking true market variability. Some products are truly seasonal or cyclical, for a host of reasons. Toys, sports equipment, and machine tools are good examples. What must be attacked is induced volatility, which makes the marketplace more variable than it really is. Once you adjust your uneven demand patterns to the true market, you can focus attention on methods to synchronize your supply with market demand.

Smoothing Original Equipment Manufacturer Demand

If your customer is a factory doing further processing on your products, then uneven sales demands probably are the result of your customer's inventory replenishment system and master schedule. I suggest you send the general manager of your customer's plant a copy of this book with the sections on smooth flow highlighted. Only if your customer's plant implements some of these ideas for smoothing flow, will you get smooth ordering. Then you can move to smooth out your flow, and in so doing, help your vendors.

Smoothing Distributor Demand

If you sell into a distribution network, either one you own or one owned by a third party, then you have some problems peculiar to this business. Erratic ordering from distributors is caused by many of the same factors that make flow in a factory erratic. Lot sizing for freight savings is one. A jumbled flow of product from central to regional to local warehouses is another. Random ordering patterns from remote warehouses that create peaks and valleys of demand on the factories is a third. A fourth problem is sales programs, such as promotions or discounts.

Each of these must be addressed separately. Lot sizing for freight savings can be reduced by shipping mixed loads of product, each in small batches, that together fill up a truck or rail car. Using more small trucks instead of a few large ones can also smooth out the flow. Truck transport rather than rail car can smooth out demand. Implementing these latter ideas could mean a freight cost penalty, but the reduced inventory and smoother flow in the factory may save more than the extra costs. Still another way to smooth flows is to design a more rational flow of product between warehouses (think of the highway versus downtown analogy for the factory). And don't forget, every warehouse is a traffic jam on the highway, sure to cause erratic flow.

Get rid of as many warehouses as possible to flow product directly from factory to end customer. This flow will always be smoother that way than when it goes through a distribution network. Don't leave inventory replenishment systems to their own devices, either. Establish a regular pattern of ordering from warehouses. Try to get this regular pattern of ordering synchronized with a regular pattern of production. Attack skewed sales patterns using the approaches suggested earlier.

Even if you implement these ideas on only a portion of the sales, you can reduce costs and smooth the demand. Just by making direct shipments to large customers in small batches you will avoid a lot of warehouse and inventory costs. In many cases the benefits will be considerable.

It really works. A division of 3M makes floppy disks. It has improved the factory process remarkably in the past two years. It has very short P times and can reliably make outstanding quality products. It is currently reducing finished goods inventories, and the ultimate objective is to make distributors forwarding agents, carrying no stock. The factory's flow will now be synchronized with the end customer demand.

The key thing to grasp is what this will do to P times. Without any warehouse stocks, product will flow from factory to customer, taking only transportation time. Products will be made to suit real demand, not to satisfy a warehouse ordering system based on forecasts. A more productive use of assets is bound to result.

HOW TO SYNCHRONIZE SUPPLY AND DEMAND

Now that we have smoothed out the demand side of finished goods, what about the supply? I talked earlier about a balanced master schedule and suggested that making products in a regular pattern would create smooth demands on all upstream work centers. It should also be obvious that this will create a smooth demand on the final assembly department as well.

This idea, shown as utopia in Figure 8-6, is called mixed model assembly for obvious reasons. Even though it may not be possible to reach the utopian condition of one product at a time, moving toward this goal obviously will pay off. Making the same mix of products each week, then each day, then each hour, as changeovers become quicker and quicker, is going in the right direction.

It should be clear how mixed model assembly provides added flexibility to satisfy urgent customer orders. If some of every product is being made every day, then an urgent order can be satisfied in a day. Producing in large batches on an infrequent cycle means that either the customer has to wait for his order or expediters have to disrupt the planned production with the attendant cost penalties. It is obvious that small batches made more often is preferable.

Handling Volume Shifts

Mix shifts are much easier to handle than volume shifts. Small lot sizes, quick changeovers, mixed model scheduling, and cross-trained workers all make the factory flexible enough to fill customer orders for specific items.

But what about a volume increase or decrease? How can this be easily accommodated? The answer lies in excess and flexible capacity. As mentioned earlier, high machine utilization results in a poor response to increased market opportunities. Excess machine capacity has to be a strategic way of capitalizing the business.

Turning capacity on or off also says that the work force must be flexible in terms of hours worked. Some companies achieve this with temporary workers, called in as needed. Still others reserve overtime to handle demand surges. Routine demand rates must be handled using normally scheduled workdays. Companies with several divisions often transfer employees from one division to another to offset varying volume requirements by division. Some companies have innovative work and pay arrangements, such as working four days per week sometimes, six days at other times, for an even amount of weekly pay.

No matter what your solution, the objective is clear. Avoid finished goods inventories and synchronize the supply rate to the demand rate. This is not always possible, to be sure, or cost effective. But make sure it's not cost effective compared to the *total* costs of decoupling with inventories, not just the accounting "hard" costs, before you rule it out.

Attack Component Stocks Next

Now that finished goods stocks have been reduced or even eliminated, it is time to attack component stocks. Pushing the assembly process toward a balanced flow means the demand on fabrication is now the same each day. The only reason for having component stocks is erratic flow caused by lot sizing, poor layout, or failure to execute in fabrication. Attacking these problems will smooth out supply, allowing component stocks to drop close to zero.

Put Your Vendors on a New Footing

Using the same philosophy, raw material and component stocks can also be reduced. This may mean that vendors have to carry excess finished goods for a while to support you, because their own supply is erratic. But now, if they follow the same route you did by smoothing out flow and balancing it to your demands, then they can push their inventories upstream, at some point to their own vendors.

As you can see, this idea of relating your own flow to your vendors and having them support you with smooth output, implies a different working relationship with vendors from the traditional, arms-length dealings we often have. Multiple sourcing of vendors and playing one off against another for price reductions will not provide the climate necessary for smooth, quick, and responsive flow. Buffer stocks and long P times will be the inevitable result of these practices.

And costs won't be lower, either. Earlier I talked about how a factory benefits from balanced flow with no inventory buffers. The same holds true for your vendors. Get them working on the same ideas and their costs will also drop, resulting in real savings that both you and they can share.

Reduce the vendor population. Treating vendors as an extension of your plant is difficult to do with a large number of vendors and few, unique part buys from each. Your demands on these parts will be difficult to level out as mix shifts change the items you need.

Many companies, recognizing this problem, have arbitrarily decided to reduce their vendor population by 75 percent or more. They now buy a wider array of things from fewer sources. Seventy-five percent is not a scientifically calculated number, but a target based on experience. They have also deliberately chosen vendors who are located for the most part within a day's travel, about 500 miles maximum. The reasons are obvious.

The company in Vienna, Austria, which I used in an earlier example, is applying this idea. Several divisions of 3M are using it, the automotive companies are pushing it, and you can see this trend accelerating throughout the country. Many people are uncomfortable with this idea. They feel there is a high risk of interrupted supplies if they are heavily committed to one vendor and that, in time, the vendor's costs will increase because of the lack of competition. These are obviously real concerns—sole sourcing of materials or purchased items is frowned upon by many companies.

But multiple sourcing has its problems too. When several vendors supply a company the same item and each one is played off against the others to get lower prices, it is hard to develop long-term, mutually beneficial relationships. It's unlikely that the suppliers will invest heavily in this company's future. On top of this, lead times often will be long, inventories high, delivery performance poor, and quality barely adequate. Some of these problems are caused simply because you can't communicate well with several thousand vendors.

It's obvious that the move to fewer suppliers must be done very carefully. Those whom you wish to work with on a more intensive basis must also want you as a customer. You need to check their capability to be a supplier much deeper than is generally called for under the traditional vendor selection method. A team from your company will need to evaluate each vendor's financial strength, production capability, quality process, engineering support, and logistics system before making any long-term commitments.

Companies that have reduced their vendor population have seen remarkable improvements in delivery reliability, quality, and even costs. A division of Square D spent $20,000 out of pocket to reduce 39 suppliers of a family of products down to 11. They saved $60,000 the first year alone in purchased costs for these same items. A Japanese saying is, "Happiness is a good, sole source vendor."

Using Flexible Capacity Throughout the Logistics Chain

I have talked about the need to smooth and synchronize flow rates. This is obviously the objective. But customers don't always want to do what is best for us. Usually they buy what *they* want.

If we have removed all artificially created uneven flows but demand still oscillates, our last alternative is to have some amount of flexible capacity. I talked about capacity contingency earlier and its role is obvious—absorb the demand oscillation that cannot be removed. This contingency must be incorporated throughout the logistics chain, in our resources as well as our vendors. Turning this contingency on or off quickly to suit the demand will still allow short, responsive P times and give the benefits previously described. Overtime, planned excess labor, and subcontractors are all ways of flexing output to suit a variable demand.

CASE STUDIES IN P TIME REDUCTION:
WHAT WORKS FOR THEM CAN WORK FOR YOU

Several published reports show what some companies have achieved with their P times. Another of Hewlett-Packard's divisions reports a reduction in WIP inventory from 22 days to 1 day. The division simultaneously reduced stockroom space by 50 percent (through smoother flow from vendors), production costs dropped 30 percent (due to better quality, smoother flows, fewer traffic jams), and it needs 22 percent less floor space to ship three times the volume as previously.

A division of Omark Industries reports a P time reduction from 21 days to 3 days, allowing it to convert from make-to-stock based on forecasts to make-to-order. Making this conversion should be the strategic objective of every company currently making to stock, no matter whether it's finished goods, intermediates, or raw materials. The Omark division obviously reduced its P:D ratio to less than 1 to 1.

General Motors, with its Saturn Project, expects to be able to manufacture and deliver one of its new compact cars just a few days after it is ordered.

There is nothing technically that will prevent most companies from making huge inroads in their traditional P times. It will take effort, to be sure. But beyond this, it will take a willingness to change traditional ways of working throughout the organization.

We have many ingrained practices that work counter to these ideas. Many of our performance measures also work counter to short, responsive lead times. This is one reason why General Motors decided to start a new corporation to build Saturn—it doesn't think it can change the mindset of traditional automobile people fast enough to get Saturn to market by the late 1980s.

I'll keep the pressure on to change your point of view in the next few chapters, especially if you're a general manager. If you aren't sold on these ideas, there's little hope for the rest of the organization. Take special note of the upcoming changes described in chapter 12. Recognize that many of the ideas in this book are simply ways to prepare for the factory of the future. Ignore them and you'll be running a factory without a future.

10 | Hurdling the Biggest Obstacle to Manufacturing Excellence: Product Variety

Excellence in manufacturing requires that you manage the variables. These variables can come from the marketplace, within the factory, or with vendors. Extreme variability is unmanageable. The best you can do is react to conditions as they occur. This suggests that all variables should be controlled. The fewer you have, the better chance of proactive management rather than reactive chaos.

One of the variables that must be managed is variety. What do I mean by variety? Any characteristic that has many configurations, for example, end products, components to make end products, processes, customers, vendors, and so forth. The greater the variety in any of these areas, the more difficult it will be to manage the business.

I am going to focus on product variety. This variable is usually managed worse than all the rest, but it has an enormous influence on the performance of a manufacturing enterprise.

A DESIGN ARCHITECTURE HELPS ATTACK PRODUCT VARIETY

The design architecture discussion from chapter 6 has elements that attack product variety. Multiple use products and standard and common component selection both address variety. But these approaches are used when creating *new* products. They do not address the problems of managing a complete product portfolio.

This is not to say that a design architecture won't influence the complete portfolio. Of course it will. But the breadth of product lines and variety of components proliferate based on many individual sales and design decisions. It is the cumulative effect of all these decisions that ends up as variety.

HOW VARIETY CREATES BURDENSOME SUPPORT COSTS

Many people feel that adding products to the product line will absorb overheads and hence generate more profits. More often, the reverse is true. As I will prove later, variety costs money.

Not only that, variety needs assets for inventories, space, and productive machinery. Instead of increasing, return on assets usually decreases.

The insidious problem with variety is it sneaks up on you. Decisions are made in isolation on a product by product or component-by-component basis. The sum total of these decisions results in too much variety to manage.

Beware the high costs of transactional volume. Costs that support variety increase after variety is added. Increased transactional volume creates a need to add an extra person in purchasing or accounts payable, for example. (Transactions are activities performed, or numbers of pieces of paper or computer action messages processed.)

While transactional volume is almost directly proportional to variety, transactions usually increase some time after new end products or components are added to the line. Hence, there is no direct tie between variety and support costs. They appear to be independent variables when they are truly dependent.

Everybody adds; nobody substracts. Another problem is that additions to product or component variety are made all the time. Deletions, however, do not receive anywhere near the same attention. Customers, sales personnel, and design engineers are always clamoring for new variations or products. Rarely is there a countervailing pressure to get rid of obsolete products. One company I know defines obsolete products as things it makes less than twice per year!

A CONCEPTUAL ANALYSIS THAT ILLUSTRATES THE VARIETY PROBLEM

The best way I know to show the variety problem is through a series of curves. These show in conceptual form the interplay of revenues, support costs, and assets. Simple calculations from these curves show cumulative profitability contribution and the effect on return on assets.

The difficulty is that it is not easy to quantify these curves. The traditional financial reporting system does not capture information useful to draw them. In fact, it does just the opposite. It shows a quite different picture that leads many companies to make poor decisions, as we shall see later.

The Curves Framework:
The Right Way to Look at ROA and Profitability

The curves I will draw will have three axes. Vertically on the left is cumulative annual percent. Vertically on the right is money, zero at the horizontal intersection, plus above, minus below.

The horizontal axis is variety. This could be the variety of end products you sell, components needed to make these products, customers, processes, or vendors. I am going to pick one of these arbitrarily—the variety of end products—to simplify the discussion. If this is not the correct one for your business, please adjust the text to suit.

Paretoize the Sales

A cumulative pareto analysis (i.e., 80/20 rule, ABC distribution) of your annual sales will be a curve similar to Figure 10-1. The first product listed on the variety axis has most annual sales, the second has second most annual sales, and the last item generates least annual sales revenues.

On a cumulative basis you will reach 100 percent of your annual sales (left axis) from the total product portfolio. You can now write in your annual sales in money terms on the right axis.

As a hypothetical example, consider a company making a variety of products with annual sales of $50 million. Its products are listed in descending order of annual revenues. The company then calculates what percentage of annual revenues each item constitutes. These per-

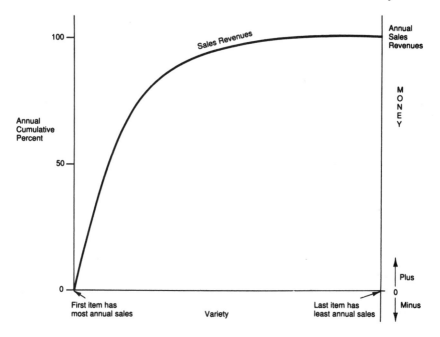

FIGURE 10-1. A Pareto View of Sales

centage values are plotted on a cumulative basis using the left-hand axis. Obviously the company will achieve 100 percent of annual revenues from 100 percent of its variety. It can now mark the right-hand axis where the curve meets as $50 million, its annual sales revenue. Hence, this one curve uses both vertical axes.

Some assemble-to-order businesses don't have this curve for their end products. Instead they have it for their major subassemblies or key ingredients. Draw the curve, then, for these items instead.

Where Are Support Costs Incurred?

First, let me define support costs. These are the total costs of the business less direct materials that are shipped to the customer (excluding scrap) and direct labor added to the product (excluding indirect activities, such as rework, material handling, changeovers, idle time or burden). Support costs include all sales activities, marketing, accounting, engineering, factory overheads and supervision, plus general and administrative costs.

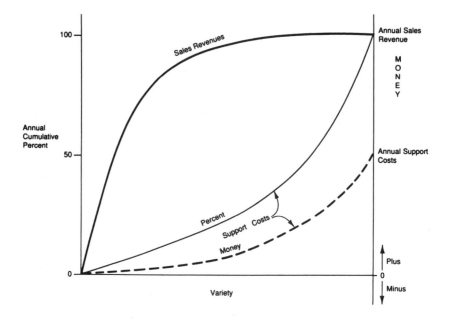

FIGURE 10-2. How Support Costs Are Incurred

The two curves added to Figure 10-1 to make Figure 10-2 show how these costs are incurred to support the product variety. The solid line uses the left axis, so it ends at the 100 percent annual cumulative percent of support costs. The dashed line is the same curve using the right axis of money.

Returning to our hypothetical company, the total of support costs as I have defined them can be engineered from accounting's records. The problem comes when trying to distribute these support costs to products.

Our hypothetical company broke its product variety into several groups; it didn't work on individual items. All department managers estimated the percentage of time their people spent on each group. Their budgets were allocated proportionally to the product groups.

The company had total support costs of $25 million. Its first group of products, the high volume, limited variety items, consumed 10 percent of these support costs. The second group consumed 15 percent, the third 20 percent, and the last group 55 percent. The company's department managers plotted these percentages cumulatively using the left axis and also plotted them using actual dollar amounts on the right axis.

Are these curves true? Ask your people if you don't believe them. Your people will tell you that the high volume/limited variety products consume far less support than the low volume/high variety products. Transactional volumes, as mentioned earlier, determine support costs, not revenue dollars. For example, the cost to process a sales order for $1 million is the same as one for $10,000. Hence, the order processing costs per dollar of revenue are far higher for the small orders.

The same relationship holds true for purchase orders, factory orders or schedules, and other support activities such as design or industrial engineering. A routing, drawing, or numerical control tape costs the same to make regardless of volume sold. Sales and marketing expenses are also linked to variety, not volume. So the curves are true, regardless of how unhappy you are with them.

How Are Assets Generated or Used?

The four key asset types are plant, equipment, inventories, and receivables. Their use or creation relative to product variety differs, so I will treat them separately.

Inventories. The ratio of inventories to sales is much lower for high volume/limited variety than low volume/high variety. Slow moving products always have low inventory turns. Hence, the curve of inventory investment relative to the product variety will be somewhere between the revenue and support cost curves. Your inventory management skills and policies determine exactly where the curve will be.

Plant. A sizable amount of your plant is used to store inventories. Hence, the use of plant to support the variety is also a curve between the revenue and support lines. It is a little closer to the revenue line than the support line.

Equipment. Machine utilization tracks fairly close to the revenue curve. The higher sales volume parts take the most time to produce. But high variety parts will need more changeovers. There probably will be more idle time caused by the problems of running these infrequently made items. Special-purpose equipment or tooling to make the high variety/low volume parts will often be idle. Hence, the curve of equipment used to support the variety will be below the revenue curve.

Receivables. I expect this curve will closely track the revenues curve. The only time this may not be the case is when the class of customers, and hence their payment habits, differ between the left side of the variety distribution and the right. For example, the high volume limited variety may go to large, successful companies with adequate cash resources, so they pay promptly. The low volume/high variety may be sold to smaller companies that are slow payers.

Total up the asset curves. Add all four curves together and you get lines similar to those shown in Figure 10-3. The solid line uses the left axis, hence it ends at 100 percent of your assets, the dashed line uses the right axis, hence it ends at your total assets in money terms.

Our hypothetical company calculated its total assets from the accounting reports. They came out to $30 million. The company allocated these assets to the product families using the judgment of key individuals. Family one needed or created 32 percent; family two, 28 percent; family three, 22 percent; and family four, 18 percent of their assets. The company plotted these numbers cumulatively using the left axis and also financially using the right axis.

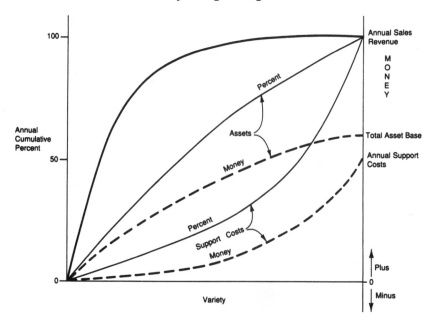

FIGURE 10-3. How Assets Are Needed

The purist will argue that the asset curves shouldn't start at zero because there are fixed and variable assets. The same can probably be said about the support costs.

But as soon as we accept that some portions of these are fixed, we stop addressing what are in truth variables. Fixed costs are not necessarily engraved in stone. We could rent out part of the building if it was empty, or even sell off a piece. We could do subcontract work for others, thereby sharing the equipment and support costs. We could even sell off some equipment if it were no longer justified.

Consider that these curves all start at zero. You'll make better decisions with this view of your business.

Calculating the True Profit Contributions

Before I convert these curves to profits, you have to be satisfied the curves are conceptually correct. If you don't agree with them, discuss their shape with your colleagues. My guess is that they will convince you the curves are fundamentally true.

I am going to calculate the cumulative profit contribution from this product variety. My formula is:

> Revenues for an item minus its direct materials and labor shipped to customers (assumed to be a curve with the same shape as the revenue curve) minus its support costs equals profits from that item.

I have plotted this cumulatively as shown in Figure 10-4.

It uses the right axis so is in money terms. It reaches a peak then decreases to the amount of money that represents your annual profits. (For purpose of this illustration, we'll presume they are positive. Of course, this curve could go negative, indicating losses.)

In our hypothetical company, the first group of items on the variety axis had annual sales of $40 million. The direct materials and labor that were shipped to customers were $20 million and the support costs $1 million. Hence, profits from this group were $19 million. This calculation was performed for each group and the cumulative total plotted. Obviously, the end result at the right axis matched the reported annual profits of $10 million.

Calculating Return on Assets

Now that I have calculated the profits, I can calculate the cumulative return on assets, also shown in Figure 10-4. The formula is simply profits divided by assets. On a cumulative basis, the curve peaks and

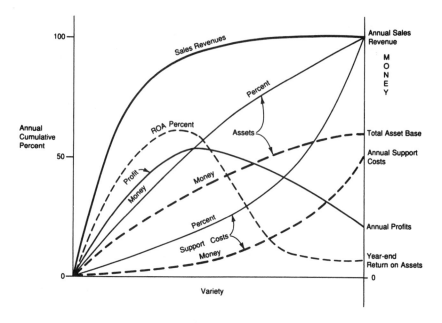

FIGURE 10-4. Cumulative Profit Contribution and Cumulative Return
on Assets

then decreases quickly. The result is obviously a percentage, and so uses the left axis. I have magnified this curve to make it visible, ignoring the numerical values of the left axis. The end point at the right is your year end return on assets.

What Do the Curves Say?

First, the last two curves of profits and ROA are simple derivatives of the first three curves. If you don't like these last two curves, go back and make sure you understand the first three. If these are conceptually correct, then the last two have to be correct.

Second, the curves say you are subsidizing some products out of the profits of others. I am sure this is not a revelation to you. The curves just show which items are subsidizing which other items.

Many companies recognize this potential problem and do a good job of support costs and assets to product families. They manage these families accordingly. However, within each family, the same problem exists. Some products subsidize others within the family. Few companies manage this part of the variety problem well.

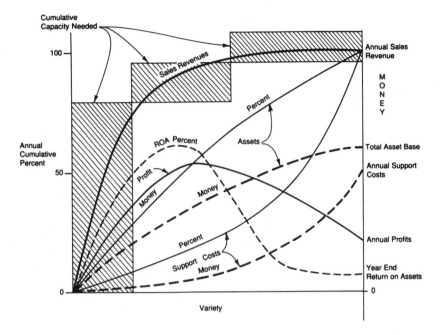

FIGURE 10-5. Capacity Consumption by Product Group

How Does Capacity Fit In?

Figure 10-5 has capacity use superimposed on the curves. The high volume grouping items consume the most capacity, the next grouping consume less, and the last grouping less still. I have used the left axis (annual cumulative percent) to draw these blocks. As you can see, I have made sure the three blocks exceed 100 percent.

Don't Make These Three Poor Business Decisions to Try to Solve Your Capacity Problem

With capacity required greater than your existing capacity, what are your choices?

1. *Work overtime.* That will obviously increase capacity but at a penalty—the overtime premium. And since this premium is incurred to produce the losing products, we lose even more money. Not a good move.
2. *Expand capacity.* Buy some new machines or extend the plant. But these additional assets will make things with negative ROA

contributions. Increasing the assets will simply make this condition worse.

3. *Sell right-hand items in lieu of left hand.* Fill up the capacity with orders for the right side products, have nothing left for the left side. But this is crazy; it's left-side products that generate your profits.

I am sure some of you are thinking that the right-side products are new products that will be tomorrow's winners. They must be subsidized and nurtured to progress to the left side. That is, of course, partially true. But also on the right side are yesterday's winners, now declining in sales and profits. Also on the right are new products that failed to perform as expected, what we often call "dogs." Making any of the above three decisions to produce these items would be a serious mistake. The question you have to answer is what proportion of the right-side products are new, needing support and subsidy, and what portion are old products on the decline or new products that failed.

Why Traditional Accounting Figures Do Not Mirror These Curves

Traditional financial systems do not associate support costs to products very well. Factory overheads are allocated to products using a burden rate based on direct labor or machine hours. This means that high volume items carry a lot of the support costs. This is not how they are incurred, if you believe the curves.

Some companies allocate purchasing, receiving, and accounts payable costs to their material purchases using a material burden rate. Again, the high volume purchases carry most of the material burden. This is also not how these are incurred.

Sales, marketing, engineering, accounting, general and administrative costs are usually gross margin deductions. There is no attempt to allocate these costs to products. What this says intuitively is these costs to support products follow the revenue curve from Figure 10-1. If you accept the transactional theory of cost creation, however, this is obviously not true.

Beware! Accounting's Figures Can Be Dangerous

Here are just a few of the ways companies can go wrong when they rely on the figures from accounting.

As decided earlier in chapter 6, company in the Midwest has a burden rate of 550 percent on its machine shop direct labor. The design engineers used to specify more expensive materials and added component variety to minimize machining. Their objective was to avoid the onerous machine shop burden. But variety increases support costs and less machining reduces the hourly base, so burden percentages constantly increased. The engineers specified more expensive materials because the accounting system said they were "reducing product costs." Actually, the reverse was true. Their current design activity focuses greatly on eliminating variety, wherever possible.

An American-based division of a European company that makes television sets transferred production of 13-inch, black-and-white TVs to Taiwan because the company was "losing money" on them. The company was "making money" on the large, multifunction, color and projection TVs, so it kept this production in America. After the transfer was complete, profits dropped!

The reason? The small black-and-white TVs needed almost no factory or technical support, but they were allocated a large share based on their production value. As the allocation made them "nonprofitable," their production was moved offshore. The large, complex units needed lots of support, but were being subsidized by the small black-and-white set production. The black-and-white sets were sent offshore, but the support costs didn't leave. Suddenly, what was thought to be profitable wasn't.

This tendency of accounting systems to shift actual cost incurrence to volume production is what makes them dangerous. Also, the gross margin deduction practice means there is no attempt to allocate the costs of a large part of the business to products. But a large portion of the costs deducted from gross margins are incurred in reality to support products. Sales, engineering, marketing, and accounting costs are all largely product-related costs. You misallocate these at your peril.

The 50 Percent Variety Figures: Are Yours This Bad?

If you split the variety axis in half, you have half the products on the left and half on the right. I will focus on the right side. Note that *50 percent of the product variety generates 5 percent annual revenues, requires 50 percent of the support costs, and uses or generates 40 percent of the assets.*

Do these numbers hold true for your business? I have deliberately been conservative to avoid biasing the discussion. My guess is that your figures are at least this bad.

What does this say about managing variety? It could be the best profit improvement program you ever started.

HOW TO START MAKING BETTER DECISIONS
BY DRAWING THESE CURVES
FOR YOUR OWN BUSINESS

The first question is: do you need to? If you believe the curves are conceptually correct, can you make good decisions based on them even if you don't have hard numbers to back them up? In most cases, the answer is yes. I am sure that a few key managers could evaluate your product variety with these curves in mind and make some good decisions of what to do. Every company has products that everyone knows are losers. Get rid of these as a start. You won't see much of a reduction in support costs but neither will you lose much revenue. You are just clearing out spurious variety and getting everyone conditioned to variety as an important variable.

Addressing the next candidates for attack will be tougher. Maybe you need some numbers to back up your decisions. This means you must attempt to draw the curves, either for the business in total or for selected product families.

The sales revenue curve is the easiest. I am sure most companies have the data to draw this.

The support curve comes next. First, break up the variety into groups, perhaps three, listing the high volume items (left side), medium volume items (middle), and low volume items (right side). You may need more groups than this to define the full variety but don't make it too complex.

Three Ways to Identify Support Costs

The support costs can be traced to product variety in one of three ways.

1. *Using management judgment.* Have the managers of every department estimate the percentage of time their people spend on the three classes of products. Use these percentages to break up each department's annual operating budget. Sum the three classes to get a view of the support cost curves.

As an example, a company in the Midwest makes standard and special valves. Company officials found that 80 percent of the industrial engineers' time was spent on special products. Their burden rate allocation method did not reflect this split, so standard valves were being overcosted and specials were undercosted. The margin contributions were erroneous, leading them to some rather questionable decisions.

They have since revised their burden allocation method. It reflects much closer the true cost incurrence. They are on their way to making much better pricing and business decisions with truer costs.

2. Taking a survey. Have everyone in the support areas keep track of their time for a period, say a week or a month, as it is spent on the various classes of products. Use this sample as the base to break up the annual budgets of each department into the product groupings you have selected. Sum each group to get the totals.

3. Count transactions. If you agree it is transaction volumes that determine true cost incurrence, count transactions within each variety grouping. This could be done manually using a sampling program or it could be done using the computer. As the computer processes each transaction, post these into the various variety groupings and sum them for each group. Use these totals to allocate the various departmental budgets.

SEVEN CHOICES FOR MANAGING VARIETY AND BOOSTING PROFITS

Now that you have the curves drawn, what choices do you have? The answer: a limited set. I will discuss each one as a standalone idea, but of course you can mix and match them to get the best result.

1. Do Nothing—But Beware of Your Vulnerability

Accept the curves and accept the need to sell losers to sell winners. This crossover sales relationship is common in many industries. Retailers often advertise loss leaders to get you into the store. They hope while you are there that you buy other items where the profit margins are good. Hence, they net out at an acceptable profit picture.

This relationship of cross-subsidizing products is also needed to support new product entry. The heavy engineering and advertising support in the early life of a product has to be paid for by the mature products in the line.

Where this course does not make sense is when the low volume/ high variety items are dogs. They are either in their decline phase or are new products that didn't catch on. There is no logical reason to supply these unless your customers buy a mixture of products, some from the left side of the variety spectrum and some from the right. If you don't have right-side products, you can't sell left side. This is often called selling a complete line and is similar to the retailing loss leader idea mentioned earlier.

But don't just accept this excuse as a reason to continue with excess variety. Force it to be proven, to ensure you are getting value from selling losing products.

Doing nothing makes you vulnerable. If you study the cumulative profit curve in Figure 10-4 you see that you make lots of profits from the left-side items and then lose profits as you move to the right.

A competitor with a limited variety line, containing left-side only products, could undercut your sales prices considerably and still end up making an acceptable profit. The low support costs guarantee good profits.

Here's an example of how this can happen. A Canadian manufacturer of bathroom fixtures and other plumbing supplies has 3,587 products in its line. Company officials are proud of every one of them. They have lost money four of the last five years.

The lowest volume item, a plumbing elbow, sells only 2 per year. The retail price is $3 each. I guarantee this item must cost at least $1,000 per year for tooling, machine changeovers, production time, inventories, stockroom space, sales catalogs, and the myriad other activities one item in the line causes.

The Koreans have recently entered this company's marketplace with a limited variety of the high volume items. They have only about 25 percent of the Canadian company's product line. The Koreans are selling items at 25 percent below the list price of the Canadian company to discount houses, such as K mart. Of course, the Canadian company accuses the Korean company of dumping. My opinion is the Korean company is making quite acceptable profit margins on its limited variety line. Because the Korean company is not cross-subsidizing losers, it can sell at considerably lower prices and still make money.

How can the Canadian company repulse this attack? One, ask for tariff protection. Two, identify its true costs and price accordingly. Three, restructure the line or its business as will be described later. Doing nothing, its current strategy, is hopeless.

2. Truncate the Line and Sell Only Profitable Items

Aggressively attack the product variety and reduce it to those items that are truly profitable. This is what turnaround artists do with losing companies. They severely prune out the losing products and lay off the associated support costs. Several companies or divisions of companies, losing money for years, have made profits three months after such an action was taken. Of course, the new boss had no "sacred cows" to keep him from making these decisions.

Two key considerations. Two key points must be emphasized if you take this choice. First, you must downsize the support costs and the asset base to make this worthwhile. Otherwise, you reduce revenues without reducing costs or assets. Second, if there are true cross-over sales, so a full line is needed, taking this action is only a temporary solution. You will lose the high volume product sales eventually, as customers reject your limited offerings.

3. Price to Suit True Costs

The curves show you the true costs associated with your products. Price accordingly.

Three reactions to changing prices. When you change prices, however, you should anticipate customer reaction beforehand. Customers can respond in three ways.

1. Acceptance. Customers accept this change and continue to buy at their past rates. This will change the shape of the revenue curve and make it closer to 70/30 than 80/20. The objective is to make the cumulative profit curve not decline but steadily increase and end higher than before.

   Some examples have been published where margins were increased 300 percent on the slow moving, right-side products. The customers accepted these changes without a murmur. It makes you wonder if your customers know more about your true costs than you do! It might also say that customers are willing to pay much more for specials than we thought in the past.

2. *Rejection of price increases.* Customers don't buy the products on the right-hand side, the ones on which you increased prices, but continue to buy the left side. In this case, market forces are truncating the line for you. Don't forget to slash the support costs and downsize the business in concert with the product sales change.

3. *Rejection completely.* Customers won't buy the low or high volume products. They want to deal with a full line producer. Your increased prices on the low volume items makes you non-competitive.

   The risk of complete rejection is what prevents many companies from adopting more aggressive pricing strategies. But you can try it out on a limited product line or slowly rearrange the pricing over several years. This will allow you to back off if you sense that complete rejection may occur.

4. Redesign to Achieve Variety Without Variety

Keep the same application variety in the marketplace without the variety of products. Remember the company we discussed in chapter 6, which made medical electronics? The company used to make both domestic and European voltage models, and it never had the right ones in stock.

The company increased standard costs when it redesigned its products to provide for a universal voltage. But its total costs reduced, and the company is now more profitable. The variety of products reduced in half and support costs have been pared.

Creating the right design architecture, as we discussed in chapter 6, fosters this approach in the early design stage. Make sure you develop such an architecture so that you don't have too great a variety coming out of design. Beware the least standard cost/most total cost trap created by traditional accounting systems.

5. Restructure the Business to Slash Support Costs

Many companies mix very wide disparities of product volumes in the same plant. But the support needs of a high volume/limited variety business are quite different from those of low volume/high variety.

Split the two businesses apart. Structure support systems uniquely for each business. The high volume/low variety business may need lots of controls, systems, and bureaucracy. The low volume/high variety business can perhaps be operated with limited information and overheads. The objective is to reduce total support costs for the full line. Profits will increase accordingly.

6. Minimize Support Costs and Assets

Many of the ideas coming from the Just-in-time (JIT) movement result in eliminating support costs, especially in the factory. Less material handling, fewer inspectors and stores people, faster changeovers, and better quality products all reduce support costs. Lower inventories and smaller production areas reduce the asset base.

The same attack can be made on engineering and marketing support. A European manufacturer of consumer electronics once charted the volume of engineering changes over the life cycle of one of its products. The manufacturer assumed the number of changes was directly related to the level of engineering support—not a bad assumption.

To the company's surprise, the level was almost constant. There was just as much activity during the dying phase of the life cycle as the beginning. The reason? The company dedicated engineers to product lines. So the engineers kept refining the design without regard to the sales volume.

Several Japanese manufacturers cut off all support to a product once it declines through a volume threshhold. They prohibit anyone—design engineer, industrial engineer, or marketing person— from working on this item, making sure in the process that it is in reality a cash cow. Instead, the companies assign these talents to new and upcoming products.

7. Don't Make Losers—Buy and Resell Them

Don't manufacture all the products you sell. Make only those where you make a profit. Buy the losers from someone else who either can make a profit on them or is selling them at a loss (I won't tell if you don't tell).

This way you become a wholesaler for the products on the right side of the curve. And you remain a full line provider of products without the costs of being a full line producer.

All you have to do is charge enough premium over the acquisition cost to pay for your buying, holding, selling, accounting, and any engineering expenses. If you made no profit on these sales, but avoided the decay of the cumulative profit line from Figure 10-4, you'd be in good shape.

SELECTING THE CORRECT CHOICE

There are a limited number of choices to handle variety. You must pick the right one for you. It is not an easy decision. Many real and imagined problems will be thrown up at you as you attack variety. Be especially wary of those who advocate "do nothing." This is a choice, but you should force every product on the right side of the variety axis to be fully justified. Keep products in the line only if you are sure that eliminating them will hurt your business. Now consider the other options of pricing to suit total costs, redesigning for application variety without product variety, or buying and reselling to solve the profit line decay.

HOW TO CONTROL FUTURE VARIETY GROWTH

As mentioned earlier, product variety grows slowly over time. It sneaks up on you because of pressures from customers, salespeople, and engineers for new things.

The reverse, reducing product variety, is an on-again, off-again thing. Product rationalization programs are started about every five years in many companies.

To stop this syndrome and apply countervailing pressure to the "expand the line" people, create a "reduce the line" program. Set up a regular review team, and schedule regular meetings (perhaps quarterly) to analyze the product offerings and recommend items for deletion. Put some teeth into this program, so candidates for deletion must be fought for to be retained.

Now select the correct choice out of the seven listed for these items. Make sure total costs prevail, not standard costs or margin contributions. With this approach you will have controlled one of the biggest enemies to manufacturing excellence that exists, product variety.

11 | Toss Out Traditional Accounting and Measure Your Way to Success

The Internal Revenue Service (IRS), The Financial Accounting Standards Board (FASB), the Securities and Exchange Commission (SEC) and shareholders require certain financial statistics about a business. These necessary measurements are usually expressed in gross terms, such as inventories, net sales, cost of sales, fixed assets, liabilities, and so forth, and make up the Profit and Loss Statement and balance sheet for a company.

A subset of these financial measurements, expressed in greater detail, is used to measure operating performance. Some of these (for example, variances and efficiency) are necessary to suit the costing system used. Others, such as machine utilization and indirect/direct labor ratios, were developed to suit views on how to control the business.

Many of these submeasures were devised for an earlier time and have not kept pace with the changing industrial environment. When direct labor was 30 or 40 percent of product costs, efficient use of labor was critical. At today's average of 7 percent, efficiency of direct labor becomes almost inconsequential.

When interest rates were low, product variety limited, and you were operating in a seller's market, inventory was of low importance. Whatever you made was eventually sold. The sudden switch to a buyer's market—due to overcapacity in many industries today and with higher quality, readily available goods from overseas—has changed the whole picture. Brand loyalty has almost disappeared as you can see from the changing variety of automobiles on our highways, television sets in our homes, and appliances that we buy. We must therefore compete with availability, quality, price, variety, and service all at the same time. We can no longer focus just on costs as we have in the past.

Hence, many of our traditional financial yardsticks, which focus only on costs, are no longer valid. But more important than that, these yardsticks tempt managers to operate poorly to get a good financial scorecard, and that is dangerous. The other competitive elements are always suboptimized in the process.

HOW INCORRECT MEASURES STIFLE PROGRESS

One of our big three automotive companies decided to pilot-test some JIT ideas in a few selected plants. It attacked changeovers to reduce batch sizes, it made product only when needed and not to a predefined schedule, it improved the layout, and so on. Inventories dropped, the company became more flexible and it was on its way to real improvement.

But these tests were all halted by the managers of the pilot plants. They were getting poor report cards from the financial system. Most plant managers are not measured on inventory levels. For those few that are, this measurement plays a very small part in their total measurement package. So the plants got little or no credit for inventory reduction, space freed up, or flexibility. But demerits *were* given for more "nonproductive" time (direct labor not producing product), indirect/direct labor ratio increases, and lower absorption of overheads.

Since people always do what you inspect, and rarely what you expect, these plant managers called a halt to the JIT process because it gave them a bad scorecard, even though this process was in the best interests of the business. The scorecard measurement system had to be changed before JIT could become a viable program throughout the corporation.

Traditional Financial Measures Are Dead Wrong

Financial people criticize high inventory levels often and vociferously. But many of their detailed measures in fact stimulate inventory growth. Purchased price variance is an easy example, where buyers often order larger quantities to get the price breaks. Rarely are buyers also measured on inventory levels—another group gets this measure, such as materials or inventory control. Other measures, such as indirect/direct ratios, stimulate increased inventories more subtly. And overhead absorption pressures are the worst measures of all.

Finance measures salespeople on dollars booked or sold. Rarely do they also measure the mix, and rarely do they care when the sales occurred in a reporting period. But selling the right mix evenly throughout time is at least as important as simply selling volume.

Several leading figures in manufacturing management have summed up this problem neatly. George Plossl once told me: "Accountants count what is easily counted, not what counts." I'm not sure I agree completely with this statement, but it certainly contains some element of truth. I would rather state it this way: "Optimized subtotals mean a suboptimized total."

Eli Goldratt says the same thing: "The sum of local optimums is not a global optimum!"

Dr. Richard E. Bird, an organizational consultant in St. Paul, MN, says: "Impeccable micro logic makes macro nonsense." These last three statements describe traditional financial reporting systems perfectly. They separate the business into small pieces and then scrutinize each piece under a microscope. The assumption is that if each piece is improved, the total will improve. Nothing could be further from the truth. What one piece does to optimize its results has a negative impact on the others. The result is that the *total* is suboptimized.

We Need Better Operating Performance Measures

If financial performance measures don't stimulate excellence in manufacturing operations, it's obvious that other, more effective measures are needed. "You can't control what you do not measure" is a famous management axiom. Hence, measurement is mandatory, the question simply is, "Of what?"

The answer to this question is a set of measures that stimulate continuous improvement in operations. Improved business results will be automatic. But this means looking at the business in total, not in detail. It's an optimized total we are after, but this may mean suboptimized subtotals to achieve this goal.

The best group to report performance against your selected measures, not create them, is finance. I say this for two reasons. One, they are remarkably regular with their reports. Once they agree on a reporting frequency, it is done with almost fanatical regularity. Second, when finance talks, everyone listens. Measures that would receive scant attention if reported by an operations person are taken seriously when finance reports them.

Make these operational measures the primary ones for your organization. Push the financial measures down in importance. Make clear to everyone it's operational excellence you are after, and that improved financial results will follow from improvements in operations.

HOW TO SELECT OPERATING MEASURES FOR YOUR COMPANY

Each business has its own unique problems and is at a certain stage of operational excellence. This suggests that operating measures should be unique to each company.

But some measures have almost universal applicability. I suggest some later, not to force them upon you and your organization, but to define the results you wish to achieve. Use them or develop your own set of operating measures that pushes you toward the correct goals.

Measuring Performance Will Be Replaced by Expecting Performance

Some of the measures I suggest will be valid early in your march to excellence but will become obsolete as you improve. They will either be too slow or built upon too static a base.

At this point you'll have to switch your attack from measurement of performance to expectation of performance, with no allowance for deviation. Examples of this are schedule performance measures and quality levels. When inventory is approaching zero, with the "oil refinery" running exactly in balance, schedule slippages and defective product cannot be tolerated. One hundred percent execution must be achieved. You will have to construct the environment necessary to accomplish this.

THREE GROUPS OF OPERATING MEASUREMENTS
THAT PAVE THE WAY TO MANUFACTURING EXCELLENCE

Three primary operating departments exist in every manufacturing company: sales, production, and design. Each has its own role to play in the move to achieving excellence. I'll separate the measures into these three groups. In all cases I'll be striving to push the total organization to achieve the mission statement from chapter 1 and look like the "oil refinery" scenario from chapter 2.

SIX MEASUREMENTS THAT RATE PRODUCTION EXCELLENCE

Measurement 1—P:D Ratio

It is obvious that reducing the P:D ratio (which, as you'll recall, is the ratio of the stacked lead time for a product to the customer's lead time) has to be a key objective for a business. Any improvement in the ratio will be positive, any deterioration, negative. It should be measured, probably quarterly, for each major product line. Your logistics planning system can easily calculate P by summing all the individually planned lead times for purchased materials and manufactured parts in each bill of material.

As discussed in chapter 3, the D time can be several things, depending on your operating philosophy. It could be what it is, as defined by realistic lead times you are currently quoting. This, of course, doesn't say whether this D is good or bad, it simply says what it is.

You could define D as what your customers want. Yet again, you could define it as that customer lead time that would give you a real, competitive edge in the marketplace.

This gives a possibility for three P:D ratios per product. I would pick one D, probably the second to start, which is what your customers want. After you attain a favorable ratio, less than 1 to 1 for this D time, I'd suggest you switch D to number 3, the time that gives you a competitive edge, and push P time reduction harder.

The value of this measure is that it involves all business functions and is therefore an integrative process. By definition it forces "total" optimization. Interfunctional teams will have to work together to make real inroads into the total P time. Design engineers, purchasing, sales, distribution, and all manufacturing disciplines will need to understand the value of a favorable P:D ratio and work together to get it.

A pictorial view of the P time buildup is shown in Figure 3-2. The impact of inventory levels is clear. The picture also shows the variabilities of lead times, a bigger influence on inventories in many cases than the lead times themselves. Use this picture to show improvements in P time. You will be able to track improvements to the business function that made them.

Remember the preferred sequence of attack on inventories from customer to final assembly to fabrication to vendor. But don't let roadblocks in any one area stop progress all along the logistics chain. Sometimes progress in the earlier manufacturing stages can clear up roadblocks in the later stages. So keep the pressure on for gains everywhere.

Measurement 2—Changeovers

The reduction in time needed to change from making one product to making another is a foundation for manufacturing excellence. The March-April 1984 edition of the *Harvard Business Review* contained an article titled "Manufacturing Performance—Pulling the Right Levers," by Larry P. Ritzman, Barry E. King, and Lee J. Krajewski. They simulated the effect of changing various operating parameters on the inventory levels of several typical plants. They state that, "the key to reducing inventory . . . is average lot size, the number of units released per order." They go on to state that, "Smaller lots reduce inventory, not only in the storeroom but also—through related cuts in cycle times, the amount of elapsed time from the release of an order until its completion—on the shop floor. In addition, they lower the number of tardy deliveries, remove unpredictable capacity bottlenecks (particularly at feeder operations), reduce the time spent waiting for free capacity, and so facilitate better customer service."

They later state that reducing lot sizes without reducing changeovers puts a heavy toll on labor productivity. So the best avenue to inventory reduction is to cut changeovers *and* lot sizes.

How to measure changeovers. Figure 11-1 shows one way of measuring changeovers. On the left are some band widths of changeover times. Three years' performance is shown for a company's program to reduce changeovers. The SMED (Single digit Minute Exchange of Dies) and OTED (One Touch Exchange of Dies) lines are clearly shown as objectives. NTED (No Touch Exchange of Dies) is not shown but it is obvious where it is. Significant progress was made between 1976 and 1977 to achieve SMED. Progress continued, although at a slower pace, to get to OTED from 1977 to 1981.

	1976	1977	1981	
>60 Min.	30%	0	0	
30–60 Min.	19%	0	0	
20–30 Min.	26%	10%	3%	
10–20 Min.	20%	12%	7%	
				SMED
5–10 Min.	5%	20%	12%	
100 Sec. − 5 Min.	0	17%	16%	
				OTED
<100 Sec.	0	41%	62%	

FIGURE 11-1. Setup Reduction

This picture is a good summary but it doesn't highlight the specific items that need improvement. This could be done in several ways: (1) with a formal Pareto (ABC) analysis of changeovers by item; (2) by identifying the bottleneck machines and reducing changeover times for the parts they make; or (3) by asking your factory supervisors, setup personnel, or direct labor workers which items take the longest to changeover. The solutions you devise for these items will probably be applicable to many others, so it is simply a matter of resources and time before you make improvements to all items.

Changeover for assemblies. Changeover is an easy term to visualize for a machine process. It is not so easy to identify in assembly. But picking and kitting jobs for assembly can be considered changeover, as can the actual conversion of assembly lines from one product to the next. The time to become proficient in assembling a new product can also be considered changeover, sometimes called learning curve. All these times need to be attacked. The picking and kitting process will be the most difficult to solve in most cases. The idea of efficiently producing one different product at a time, preferably on a mixed model line, is still your objective.

Changeover for transportation. For transportation into and out of the factory, changeover time is also a difficult concept to visualize. But if you understand that the objective is small lots flowing regularly and often, as the refinery example shows, then changeover becomes easier to define as "the time and effort required to make a pickup or delivery." Reduce this by routing trucks on a regular schedule to vendors that are close together and fill each truck with many small lots of different parts. Doing the same for outbound deliveries to customers keeps the refinery analogy intact throughout the logistics process.

A company in Holland uses contract truckers for all its incoming and outgoing freight. It used to use a variety of common carriers, based on their availability, type of freight, location, and so forth. It recently contracted with one trucking company for all its freight, and has defined a standard routing and frequency. Freight costs have dropped at the same time as smaller lots are received and shipped. The company is gaining in both directions, inventory and costs of transportation. It's obvious how reducing the number of vendors and having them all within a limited distance from the plant can foster this kind of gain.

Measure changeovers or batch sizes on a regular schedule, probably quarterly. Force them both down. The benefits will be phenomenal.

Measurement 3—Schedule Adherence

In chapter 2 I defined two of the five required elements of operational excellence as flow rate control and sequence control. I also stated that making just enough of just the right things is more productive than making too much, too little, or the wrong things, efficiently.

A simple way to measure schedule adherence. Frank Gue of Westinghouse, Canada, in his book, *Increased Profits Through Better Control of Work In Process,* Renton Publishing Co., 1980, defined a remarkably easy but effective way of measuring both flow rate and sequence control. His measure does not require any more data than you already have on hand. But it shows graphically what utopia is and where every work center or vendor you choose to measure is operating in relation to utopia.

Figure 11-2 shows three examples of Gue's charts. Time is marked off horizontally in weeks or days, the midpoint established as on time, periods to the left late, to the right early. Vertically, the axis represents output, expressed in product flow terms, such as units, hours, dollars, and so forth.

Collect all production reports for a resource for the measurement period: day, week, or month. Compare the actual completion dates of production to the scheduled completion dates. Calculate their periods early, late, or on time.

Plot these results on the chart. Plot the amount of production vertically and adherence to schedule horizontally. A histogram for each resource will result.

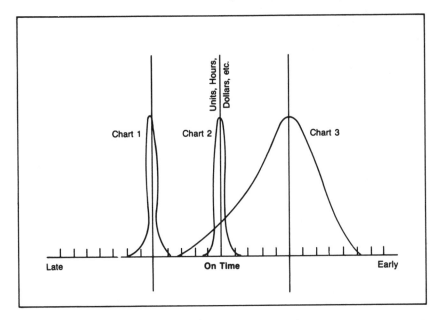

FIGURE 11-2. Monitoring Schedule Adherence

Three typical charts are shown on Figure 11-2. Chart 1 defines a resource late on the average, but consistantly late on all items. The fact that it is late on the average tells you that this resource or an upstream resource is lacking capacity, the ability to flow product. The only way to move the average to be on time is to add capacity to the bottleneck. If the chart shows that the resource is late by the same amount of time week after week, then a temporary addition of capacity is all that is needed. Once the average is on time, removing the temporary capacity will cause the average to stay on time.

If the average is slowly moving left, it means there are both permanent and temporary shortfalls in capacity. Adding capacity permanently will stabilize the drift to the left. Adding more capacity temporarily will push it to on time on the average.

The work center shown in Chart 1 has good sequence control. All items are being completed late by about the same amount of time.

Compare this to Chart 3. The average is early, which means it is flowing too much, but at the same time it is producing some things too late, as shown by the tail to the left of on time. Late orders in a work center often trigger adding capacity, through overtime, to solve the problem. But this work center doesn't need overtime, it needs undertime! It doesn't have a flow rate problem, it has a sequence problem. Solving this through better control of queues, sequencing decisions, or better sequence at the upstream work centers will solve the late order problem quickly, without the added penalties of overtime and the excess inventory that overtime would create.

Chart 2 is obviously utopia. Its average is on time, indicating flow rate balance and its sequence control is excellent.

Some companies use these charts with the time line based on order starts, not order completions. They feel the additional time this gives them to react to deviations is worth this change.

Stratify the work centers and get senior management's attention. Gue goes on to suggest a way of focusing senior plant management's attention on the critical problems these charts depict. He suggests two additional charts, shown in Figure 11-3.

The top chart, A, is a list of the measured work centers with the most late on the average at the top, the most early on the bottom. This identifies those work centers that are not currently flowing enough and those that are flowing way too much.

Chart A shows a situation that obviously needs a flow rate balancing attack. Either resources must be taken from work centers low down on the list and given to those at the top, or work has to be taken from the top work centers and given to those on the bottom.

Chart B shows each work center's sequence control. The one at the top has the widest spread between early and late, the one on the bottom, the narrowest.

A sequence attack is needed on the top work centers. There are three ways to narrow the band between early and late: (1) Force better adherence to schedule through a more organized queue, (2) improve scheduling control, or (3) give production people less leeway over what to run next.

Measure vendors too. Vendors can and should be measured exactly the same way. The only difficulty is determining what the output unit of measure should be. Pick units, dollars, lots, or whatever feels good, and don't get hung up on whether one unit contains more work than another, or this lot contains more parts than that one. As long as you are consistent, the real information you are after—flow and sequence control—will come through.

A large manufacturer of computers for the defense industry cre-

A. FLOW RATE PROBLEMS

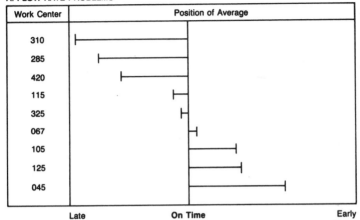

B. SEQUENCE PROBLEMS

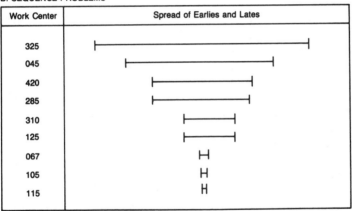

FIGURE 11-3. Identifying Flow Rate and Sequence Problems

ated these charts for some key vendors. Its capacity unit of measure is lots. The manufacturer showed its charts to pertinent vendors, whose reactions were, "I can see our performance clearly now. Your previous measure, percent on time, meant nothing to us. This is so crystal clear, and the objective obvious."

The company was astute enough to realize that it was partly to blame for the poor performance of its vendors. Crisis expediting, which it was good at, pulled some things in early, but as the vendor's capacity wasn't increased, it was accomplished by penalizing some other items. Hence the spread of early and late was wide. The company also realized that it pressured some vendors to accept more work, even though those vendors were currently overloaded. This forced deliveries late on the average. The company is now working with these vendors on a program to solve these previously hidden problems. They both get a regular scorecard, the chart, prepared monthly in this case, to show them how they are progressing.

Address schedule adherence problems in sequence. Be careful of the sequence you choose to solve these problems. If there are three resources in a row, the first one feeding the second and the second feeding the third, then it is obvious that the work centers which follow cannot perform much better than their feeding resources. If the first work center is late on the average or has a wide spread between early and late, then the following work centers won't be able to do much better unless they have large queues or a sloppy scheduling system. Solve the first work center's problems and the others will be quickly solvable too.

Schedule performance must switch to schedule expectation. This type of measurement requires a schedule to measure actual performance against. As such it is valid for a relatively slow planning and scheduling process, or what is termed a push system. Schedules are issued and we expect each resource to execute to the schedule.

As the velocity of material flow increases, though, schedules become more dynamic. They must respond to what is needed to ship a product ordered just a few hours or days ago. We could migrate to daily schedules and daily performance monitoring, but at some point even this becomes too slow and cumbersome.

Furthermore, as the velocity of material increases, the discipline of factory performance must increase. This is when you have to switch from performance measures to performance expectation. One hundred percent schedule execution must become the norm—anything less will disrupt the product flow.

The implications of this are first, to eradicate anything that can interfere with schedule execution. Quality levels, machine maintenance, operator training, and vendor relationships must all be good enough to preclude most interruptions. Second, you must be able to handle the few remaining scheduling deviations quickly by adjusting capacity to recover from any losses or by having production flexibility or redundancy to produce the needed items. One hundred percent schedule execution has to be expected and delivered. You will have to create the environment to make this possible if you wish to reach this level of excellence.

Measurement 4—Distance Traveled

Communications problems increase as the square of the distance between the speaker and the listener. My experiences supplying the component needs of seven affiliate plants around the world from one central plant taught me this clearly. The distance materials or products travel also has a large influence on P times, whether inside the plant, in the vendor cycle, or the distribution cycle.

Hence, a simple and effective measure of the quality of the logistics process is the distance materials travel as they are produced, procured, and distributed. The shorter the distance, the better the logistics and the shorter the P time. This is a simple take-off on the downtown and highway analogy used earlier. The shortest distance between two points is a straight line. To reduce distance traveled you must force materials to travel in a straight line.

One of my clients in California calculated that his major raw material, a forging weighing 2,000 pounds, traveled 2½ to 3 miles from receipt to shipment as an end product. And this travel was all done inside the plant. Worse still, when I gave a seminar in Windsor Locks, CT, one of the attendees told me he calculated that the distance their main raw material, a large casting, traveled within the plant was 19 miles!

Without any information about either of these company's products or processes, I am sure your reaction is, that's ridiculous. For product to travel several miles within an average size plant—and these plants were, if anything, smaller than average—the product is obviously moving between many resources in a very complex fashion. They have downtown city centers in their factories, not highways. And travel in downtown traffic is slow, with many stops and starts. Hence, P time is long.

Check vendor distances too. If we go outside the factory and measure the distance raw materials travel from vendors to our receiving dock, we get similar ideas. Long distances mean long times. Coordination, communication, and cooperation are more difficult and more prone to error.

As I pointed out earlier, my client in Vienna is transferring his purchases to vendors within 300 miles of his plant not only to reduce P time but also—and perhaps more important—to improve his relationships with vendors. Several 3M divisions are doing the same.

Be on the lookout for stoplights. The number of stops and starts products make as they are processed is also interesting. Traveling 2½ to 3 miles on a highway is one thing. On a road full of stoplights it is something else again. But generally, don't go to the additional effort of counting these interruptions in flow. You'll make the measurement too complex. Just concentrate on distance—you know shorter is better. By reducing the distance, you'll have to reduce the stops and starts. And reducing the distance will pay off, not only in shorter P times, but also in your ability to better coordinate your plans with your vendors—and possibly even your customers—for the benefit of all.

Measurement 5—Order Changes

Three kinds of orders are used routinely by manufacturers. Customers' needs are defined on sales orders. Your wants from vendors are defined on purchase orders. And production is authorized using a shop order or schedule.

Each of these orders defines three things: an item, a quantity of that item, and a schedule date for the item. In the case of a customer order, the item is one of your finished goods or possibly a spare part. With purchase orders, the item is your raw material or component. And shop orders could cover any phase in the product's process.

The term "order" suggests a degree of solidity or stability that in reality doesn't exist. Original orders are modified all the time by change orders. The number of change orders you get, for any of these three types of orders, tells you a lot about the effectiveness of your planning and control system. And indirectly, it tells you about your progress with both P and D times.

Figure 11-4 shows the triple forecast error problem again, but with the added concept of "horizon of stability." It shows that orders placed within a certain lead time are remarkably stable, and forecast error is almost nonexistent. Place or book orders beyond this point and forecast error will create change.

Here's an example of how this can happen. A division of a large company in Europe has six assembly plants, and they all get their metal and plastic components from one internal parts manufacturing plant. By corporate edict, the assembly plants must place orders for metal and plastic parts using a lead time of 14 weeks. Change

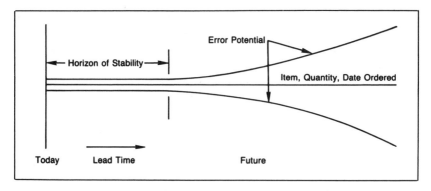

FIGURE 11-4. Horizon of Stability

orders are frequent, and delivery performance to the latest change order is poor.

An analysis was made of the number of changes during the 14-week life of their orders. Up to 7 weeks after order placement, almost no changes were made. After 7 weeks, revisions were frequent and often drastic. The result was open warfare between the assembly plants and the component plant, each blaming the other for the problem. The problem was in reality the lead time—D for the component plant, a part of P for the assembly plants. The lead time was too long. It was past the point of stability, well documented in this case as 7 weeks.

No solution other than reducing lead time can solve this problem. But instead, the supplying plant developed more complex systems and tried to resist changes after orders were placed. Inevitably, these "solutions" failed. Progress will only happen—and then it will be dramatic—when lead times are reduced so that they fall within the horizon of stability.

Measure changes per type of order placed or booked. After reducing its P time, one company found change orders to its purchase orders dropped from an average of 6 per line item ordered to 1 per 100 line items ordered. I'll leave you to figure out how much reduction in non-value added waste (NVAW) this company and the vendors gained as a result.

Measurement 6—Data Quality

Logistics systems live or die based on the quality of the information they use. Stockroom records, bills of materials, routings, purchase and customer orders, and work-in-process inventories are all records that must be accurate. Sad to say, many of these records contain significant amounts of error.

It should be obvious that as inventories drop and the company tries to be more flexible, data quality must improve. What may be good enough in a company with high stocks and slow reaction will be completely inadequate with low stocks and fast reaction.

Contrary to some opinions, this is not a systems problem. It's a management problem. Only when you decide you need accurate information to operate the business will you get it.

Don't try to inspect quality into the data. Many people feel the solution to poor information is frequent checking and reconciliation programs. Nothing could be further from the truth. You can't inspect quality into products, and you can't inspect quality into information.

Even though I am sure you agree with these statements, this is not the way industry behaves. Physical inventories are common, and so is cycle counting (that is, counting a few parts every day until you cycle through the inventory in some time period). Both these methods attempt to inspect quality into the data. They don't work.

How to improve data. The solution is the same as with product quality, which is to improve the production process so it can make only quality products. Improve the information process so it can provide only good quality information.

What this means is that any checking of inventories or other record types must be done with the objective of identifying the process problems causing bad information. Eliminate the process problems and the records become good. At some point, eliminate all data quality checking and devise self-checking methods within the process itself.

SIX MEASUREMENTS THAT GAUGE PRODUCT DESIGN

Measurement 1—Total Number of Different Parts

Count the total number of different raw materials, purchased parts, components, subassemblies, and major assemblies in a design. Try to reduce this number. The higher the number, the more complex the logistical process will be for this product and hence, the higher the risk this product will not succeed in the marketplace.

Count the total number of items on the data base. Create a program to reduce this number by standardization, elimination of unwanted parts, and so forth.

General Electric (G.E.) subscribes completely to the transactional theory of cost generation. G.E. officials believe many items contribute to transactional volume, but a major one is part variety.

One G.E. business gave incentives to its draftsmen if they could eliminate active part numbers in their system. In two months they had eliminated over 1,000 part numbers, which translated into a reduction of 39,000 annual shop releases and resulted in 125,000 fewer pieces being ordered, fabricated, stocked and inventoried. They believe that is productivity.

Measurement 2—Percent Standard, Common and Unique

This measure comes directly from the design architecture discussion in chapter 6. Measure the ratio of standard, common, and unique parts in every new design and measure them in the total data base of parts. Push for most standard and second most common, and try to limit uniques to where they are absolutely necessary for performance or differentiation.

This percentage measurement could conflict with measurement 1, the total number of parts. To reduce the total number may require more specialized parts. Be wary of which of these two takes precedence. If the unique parts have short lead times you will benefit from fewer parts. If they have long lead times you will be better off with more total parts.

Measurement 3—Number of Different Processes Needed

Just as with parts, the more processes you have, the more logistically complex your business. Measure the number of different processes in a design and the number required to make all designs. Push to reduce both.

Measurement 4—P Time for Critical Path

Measure the expected P time for a given design. See how you can reduce it with redesign.

If the consumer electronics company described in chapter 6 had measured the P time of its design for stereo equipment, it would never have released this product, no matter how good its performance quality. It would have been patently obvious what troubles this design would cause.

Measurement 5—Position of Variability

Measure the points at which the design progresses from common raw materials and components into variable end products. In other words, define the shape of the mushroom in Figure 3-7.

Do this using a time indented bill of material. Count the number of different items at various time stations in the product buildup. Work to reduce the number early in the process. Push the variability as late as possible.

Measurement 6—Number of BOM Levels

The more levels in a bill of material, the more complex the logistics will be. Utopia is a single level bill of material (BOM).

Design the product and the process to avoid intermediate stages. This is easiest to achieve during the assembly stages. Flow parts into subassembly lines, flow subassemblies into the final assembly line. If you have to define these sublevel products, make sure they are transient, that is, get built and consumed almost immediately. That way the logistics system will ignore them and complexity will be avoided.

FIVE MEASURES FOR SALES

Measurement 1—Forecast Quality

It's a rare company that doesn't use forecasts to help plan its business. P:D ratios greater than 1 to 1 demand that forecasts be used to plan intermediate inventories. It's amazing to me how many companies make forecasts and how few measure their accuracy. Measuring forecast accuracy is not intended to penalize the people who make poor forecasts, but to learn how to improve the forecasts themselves. If you can't improve them, then use forecast accuracy measurements to establish contingency plans. These must buffer your customers from the quantified poor predictive nature of the business.

You can measure forecast quality using the same concepts that underlie measuring schedule adherence. What you must measure about forecasts, however, are these: (1) your ability to predict volume, that is, the sum total of all the items forecasted; and (2) your ability to predict the mix, the specific item accuracy.

A take-off from Frank Gue's schedule performance charts is shown in Figure 11-5. The horizontal axis in this case is percent of forecast and the vertical axis is a measure of products sold (dollars, units, tons, and so forth).

At the end of a forecast period, measure actual product sales against the individual product forecasts. Plot the results on the chart.

As an example, let's say two products—Product A and Product B—make up a family. Both products were forecasted to sell 100 units.

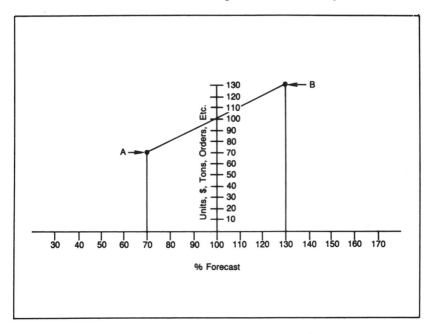

FIGURE 11-5. Measuring Sales Forecast—Volume and Mix

At the end of the forecast period, Product A came in as 70 units sold, product B as 130 units sold.

In Figure 11-6, Product A has been plotted as 70 percent forecast with 70 units; product B as 130 percent with 130 units. The average is in the middle, exactly at 100.

What does this trapezoid tell you? It says we can forecast the average sales (volume) very well. We are not too good at mix. Utopia would be a straight line at the 100 percent point.

Make the same chart for your own product forecasts. If the position of the average is on the left, it says you are underselling the plan. If it's to the right, you are overselling the plan. The spread between the percentages tells how well you can predict the mix.

Which forecast to measure? If you have a P time of six months, purchase orders are obviously being placed based on your prediction six months out. As time goes by, you can use shorter range forecasts to initiate production of certain items. The one-month-out prediction is probably used to decide what to assemble.

Let's assume monthly forecast revisions. Which of the six forecasts for this product should you measure? The answer depends largely on the "shape" of your designs. If they are mushrooms, then *volume* is your critical forecast characteristic for most of the production process. Volume is what drives the purchase or production of common raw materials and parts. The mix forecast is important only when you finally process these standard items into the various end products.

However, if variety starts much earlier in the product's process, then both *volume and mix* become important forecast variables. Hence, you need to measure them both several forecast periods before you make the actual sales.

Measurement 2—Customer Satisfaction

Customers build up their perception of your performance from many tangible and intangible contacts. To satisfy them, your performance must be good in these three key areas: quality, delivery, and technical support.

Customers are also subject to the recency syndrome, that is, their last experience and whether it went well or poorly. Their objectivity is clouded by that last experience, no matter how good—or poor—your products and services have been overall.

To offset this, establish a routine measurement system of your performance vis-à-vis your customers. Publish the result and show it to them, so they can see what your long-term performance is in these three key areas. At the same time, include some goals that show where you would like to be.

Your quality levels could be determined from internal measures of quality or from warranty or service calls. You could measure delivery performance using the Frank Gue schedule performance charts, based either on customer promise date or request date. Technical support is more a qualitative measure than quantitative. Perhaps you can get this kind of information from a survey of your customers.

How to use snake curves to make sure you and your customers are on the same wavelength. One method of learning about your customer's requirements from your logistical system, as well as your performance in relation to those requirements, has been dubbed "snake curves." An example is shown in Figure 11-6.

Here's how to use snake curves. First, list various logistical characteristics on the left. I have provided a few to give you the idea. Across the top you have two scales, importance and rating. Importance means how important your customers feel that characteristic is for the success of their business. Rating means how they rate your per-

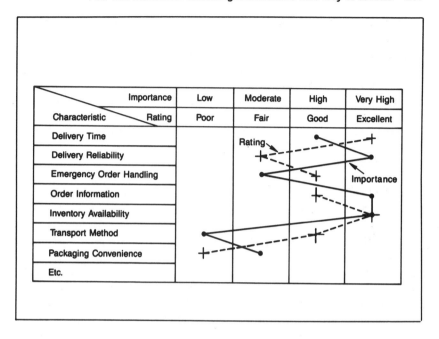

Characteristic	Importance Rating	Low Poor	Moderate Fair	High Good	Very High Excellent
Delivery Time					
Delivery Reliability					
Emergency Order Handling					
Order Information					
Inventory Availability					
Transport Method					
Packaging Convenience					
Etc.					

FIGURE 11-6. Assessing Marketing Support

formance for each characteristic. The two curves shown, dubbed snakes, show a considerable disparity between what customers feel is important and how you are actually performing.

Both these curves should come from a formal survey of your customers, which can be made by your own marketing group or an outside firm or university. An outside group could also extend the survey to include your competitors. That way you can see where your competition is strong and weak, where you are strong and weak, and how these curves match the customers' needs. An action plan to superimpose your ratings curve on the customers' importance curve would really pay off.

The curves may surprise you. Many sales and marketing people feel they know their customers' needs. They don't need to survey them formally. But more than a few have been shocked at the findings of a formal survey. Here's just one example. The marketing department of a carpet manufacturer made this survey recently and found out that many products it was offering from stock didn't need this level of service. Customers were quite willing to wait several weeks for these products, so these carpet styles could actually be made to order. The reverse was also true. Survey results showed that customers wanted off-the-shelf products the company thought could be made to order.

The marketing people have completely revised their logistical support in line with customers' requirements. The beneficial effects of synchronizing these two curves are already being realized.

Measurement 3—Demand Stability

To achieve the utopian factory, as described using the oil refinery analogy, you want to produce at a level rate. Some volume flexibility is possible in every plant, depending on how much excess capacity is available and how fast you can turn that capacity on or off.

Extreme volatility in volume demand is impossible or uneconomical to handle. Decoupling methods, such as finished goods inventories, are often interjected between demand and supply. But these inventories are NVAW because of unstable demand. The better alternative is to stabilize demand as far as possible. Some amount of instability in demand rates is caused by internal policies. The periodic sales targets and billing terms, described in chapter 9, as well as discount structures and erratic advertising, all amplify the inherent demand variability. Stopping this amplification is one reason to measure demand stability. Constructing plans to influence the market to dampen its variability or developing procedures to handle the inherent variability is the other.

Five costly problems caused by unstable demand. Unstable demand always causes disruption. Disruption always ends up costing you money. Here are five ways this can happen.

Invoice errors. Many companies have an end-of-the-month, end-of-the-quarter, or end-of-the-year syndrome. Superhuman efforts at these times push more product out the door. But these superhuman efforts often overload the resources. Mistakes are easily made, and these mistakes end up as billing errors. Now the cycle of customer complaints, salespeople verifying these complaints, credit memos, and the like begins. All this is NVAW.

Excess costs. The end of the period push is often accomplished via extreme factory overtime and air freight, either to get parts and materials in or ship products out. Both cost money. At the same time, people are often so worn out after this hectic period that efficiency drops and idle time increases at the beginning of the next period. Hence, more costs.

Excess and unbalanced inventories. Demand variability is very often compensated for by extra inventories of finished goods or work-in-process. The Northeast computer manufacturer described in chapter 9 that sold 20 percent of the quarter's revenues in month 1, 30 percent in month 2, and 50 percent in month 3 certainly took this tack.

But extra inventories are valid only if you can forecast the need for them with a high degree of accuracy. With a dynamically variable demand, this would be extremely unlikely. So excess inventories and the wrong inventories will be inevitable.

Poor service. A key part of customer satisfaction is delivery on-time, when the customer needs the product. Uneven demand, by definition, will result in poor delivery performance. The forecast error just mentioned will cause it. So will an overloaded factory at the end of the period, because there is no capacity available to take care of additional customer demands.

Factory dynamics. A factory is most efficient and effective with stability of output. If extreme changes are made to the scheduled rate of output, the cost will be high due to expediting, poor quality, shortages, and general disruption.

How to measure demand stability. Let's define the objective first: "If you want to make daily, you must sell daily." Demand stability measures help you understand your demand volatility. They also show you how to stabilize it.

For a make-to-stock company, the order entry rate should be measured each day. This information should be charted day by day as well as cumulatively for the month or quarter. Doing this by sales territory or even by salesperson may help to locate those causing peaks and valleys of demand. Work with them to see how to smooth the flow of orders.

For a make- or assemble-to-order company, it's not so much the order entry rate that needs to be level as the requested delivery dates to customers. Available-to-promise techniques in master scheduling can make the promised delivery dates level, but now you may not be fulfilling the customers' requests.

Measure the order request rates for make- or assemble-to-order products each day or week as well as cumulatively for the period.

Again, break these numbers down by territory, salesperson, or even type of product to try to locate the reason for instability. Develop action plans to level the order requests wherever possible.

Measurement 4—Product Variety

Chapter 10 was devoted to the whole concept of variety as a variable you must manage. Routinely report the number of end products you offer, maybe grouped into families, if you are a make-to-stock company. Do the same for major assemblies if you are an assemble-to-order business. Challenge any increase and push for a decrease.

Measurement 5—Master Schedule Changes

The master schedule and its role was described in chapter 4. Changes to these numbers explode down through the details of the logistics system, causing many changes and costing lots of money.

Identify how many volume or mix changes are requested by sales for each schedule planning period. Calculate this number within any time fences that have been set, for example, within the first six weeks, the next eight weeks, and beyond. Try to minimize the number of changes through better forecasting and planning. As P times reduce, reduce the time fences.

HOW TO INSTILL IN YOUR BUSINESS
AN UNSTOPPABLE DRIVE FOR IMPROVEMENT

It's easy to see the direct benefits that significant progress in these measures will bring. Improved quality, lower inventories, higher real flexibility to change, more productive assets, a more predictable process, and improved customer satisfaction will all result. To a far greater degree, you will be able to achieve every business's three objectives: increasing throughput, reducing, costs and generating cash.

Indirect benefits may be tougher to identify but could be of more value. Improved machine maintenance, better-skilled labor, unplugged bottlenecks, logistically friendly designs, and a smoother demand stream will all make your business flow smoothly, like an oil refinery. But maybe the biggest benefit will be an increase in *visible management* where problems are easily seen. With the process streamlined into a constant flow of product, problems will create an interrupted flow, as visible as a traffic jam on a highway. Fast response to remove and eliminate these problems will force the organization toward higher and higher levels of competence.

This process is shown in Figure 11-7. Start the process by locating a problem. Do this by reducing inventory until a rock is visible or find a problem using the measurement system. Identify the cause of the problem, remove it, and standardize the solution into the operation.

Standardization Is the Key to Solving Problems Forever

How many times have you or your people solved a problem only to have it recur a few months later? This cycle of solving problems and returning problems is what standardization attacks.

What do I mean by standardization? Formalizing every procedure and process that eliminates the chance of recurring problems. For example, standardizing the changeover process of a machine ensures it is always done the same way, so product quality will be repeatable and the changeover time short.

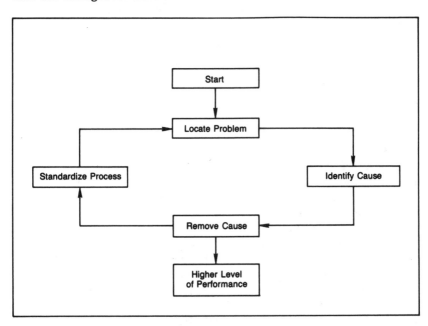

FIGURE 11-7. The Continuous Cycle of Improvement

This does not mean you resist improvements or changes. Standardization provides the foundation from which to progress. If you don't know where you are, how can you get to where you want to be?

Henry Ford, in his book, *Today and Tomorrow*, Doubleday, Page and Co., 1926, constantly affirmed the value of standardizing the variables. His progress was largely the result of finding a better way, institutionalizing this better way, and then constantly trying to find an even better way. And while he was trying to find a better way, standardization ensured there was no risk of regression or of working on the wrong problem.

Standardization is a step that is poorly done in many companies, probably because they create separate staff functions to do it. Instead, have each department develop its own standards. In many cases, don't even type them. Handwritten instructions often help people realize the transient nature of standards. They are more likely to add special notes and revise the standard with handwritten documentation if a new way is found.

Ensure that supervisors check periodically to make sure the standard method is being used. This follow-up will show that the company is serious about finding and using a better way. Make sure supervisors or fellow workers also train new people to use the standard method. This way the experience of previous workers in a job is not lost. The standard procedure will reaffirm the correct method and help answer any questions.

The Improvement Cycle Will Lead You to Business Excellence

It is obvious that every business function can improve its operation to benefit the business in total. The lake and rocks analogy clearly illustrates this concept.

The manufacturer's mission statement, to flow product from vendor to customer, defines for all functions where they must focus their attention—that is, on logistics.

Have each department start its own cycles of improvement. As soon as one problem is solved and institutionalized, move on to the next. Each cycle will push you to a higher and higher level of performance. At no time will this cycle stop generating benefits far in excess of its costs. You will have succeeded in instilling into the organization a drive for improvement that is unstoppable.

12 | How to Prepare Today for the Factory of Tomorrow

More changes will occur in industry in the next 10 years than have occurred in the past 100. Technology will push these operational changes on us—the first time technology has been the driving force since mass production. Some companies will be winners and others losers during this period of change. The winners will be those who understand and apply this new technology early and effectively.

Some of these technological changes are already here in rudimentary form, others are under active development, and still others are no more than concepts and ideas. Using today's knowledge, we can envision how the factory of the future will operate:

> A design engineer draws a part on a video screen and enters the specifications into a clearinghouse for bid. Seconds later, an acceptable bid is made by one or more local facilities, and an electronic contract is formed. Two hundred units of the specified part are delivered the next week having been made at a facility with just the right machines and job schedule to minimize their cost.

This futuristic view of a factory is provided by Richard Rumelt in his paper, "The Electronic Reorganization of Industry," presented at the Global Strategic Management in 80s Conference, London 1981. Notice the time elements, especially, "seconds" and "the next week." It may sound like science fiction, but there is no doubt that we are moving in this direction.

In the future an order will be received electronically and designed uniquely for the customer on a computer aided design (CAD) system.

The process for making the item will be automatically created on a computer aided process planning (CAPP) system which takes into account the resource capacities and current workloads. Next, instructions will be given to the automated storage and retrieval system (ASRS) to pick materials and tools. An automated guided vehicle (AGV) will move these to the correct process steps, and then robotics and direct numerical control (DNC) machines will produce the products and the AGV will deliver them to the shipping dock. Artificial intelligence (AI) will tie all these elements together and make adjustments for any conflicts. All of these steps will be automated, under direct computer control. Data will move from customer to supplier and through their respective facilities smoothly and electronically, without the stops and starts caused by human intervention that we experience today.

We are already approaching this scenario in several process industries, including refineries and breweries. Fabrication and assembly plants are still too complex for today's level of technology. But technological developments, together with the ideas advocated throughout this book to simplify the manufacturing process, will make it possible to achieve Rumelt's futuristic view even in these industries, and probably within ten years.

The relationship between the process industry and discrete manufacturers in their march toward complete automation is shown in Figure 12-1. Components refers to pieces of systems, for example, a computer numerically controlled (CNC) machine tool, purchasing system, or general ledger system.

Manufacturing resource planning (MRP), computer aided design (CAD), and accounting systems can be thought of as stand alone systems. Integrated control systems hook different systems together, for example, MRP plus accounting becomes MRP II. This is why MRP II is considered to be a comprehensive planning and control method-

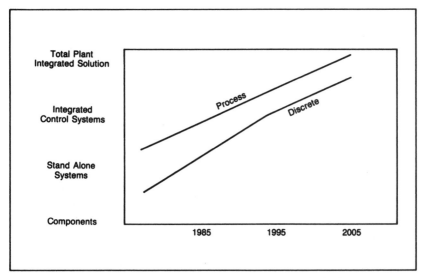

Total Plant
Integrated Solution

Integrated
Control Systems

Stand Alone
Systems

Components

1985   1995   2005

Source: U.S. Office of Technology Assessement

FIGURE 12-1. Evolution of Integration

ology, pictured schematically in Figure 4-1. Total plant integrated solution refers to Rumelt's concept of the factory of the future, in which all control data in the company is linked together, regardless of where it is located and how it is used.

As you can see, the process industry is ahead in applying these systems. If you want to see how a discrete factory will operate in ten years, go see a process industry today.

The gap between these two industries will narrow as new technology comes on stream to help the discrete manufacturer. Developing the new technology will be the easy task. Implementing and managing it will be the tough one.

SIX WAYS TO PREPARE NOW FOR A FACTORY WITH A FUTURE

We can visualize tomorrow's factories today. Admittedly, some parts of the picture are hazy, but the broad outlines are clear and some pieces well defined.

You can do several things right now to lay the groundwork for the future. They are fundamentally valid, and the beauty of them is that they will improve your operation today, generating the funds and initiative to apply tomorrow's technology quickly, as it evolves. This will put you ahead of the pack, a critical place to be as the pace of worldwide competition intensifies.

1. Simplify the Environment to Cope Better with Complexity

I make no apologies for repeating here a theme I've emphasized throughout the book: simplify. Far too many manufacturers have let their businesses become too complex. Products, markets, factories, vendors, and distribution patterns create a maze of activities that defies understanding. Managers bounce from problem to problem in this environment, and management by crisis (MBC) reigns supreme.

Business is complex today, and is getting more complex. Product life cycles are shortening because of quickly changing technology and broader competition. Traditionally domestic markets are being attacked by overseas competition. Government regulations, tax changes or proposals, and interest rate fluctuations all combine to make the job of a senior manager in industry more and more difficult. These external influences are not going to go away. If anything, they are going to accelerate. A senior manager's job is not going to get any easier.

The only way to combat this increasing external complexity is with internal simplification. You *can* control your product designs, you *can* control your factory's operations, you *can* control your vendors, you *can* control your distribution and sales networks. As you simplify internally, you'll free up people resources that can be used to better manage the external complexity.

Use the river and rocks analogy from Figures 2-7 and 2-9 or the P:D ratio of Figure 3-3 to push you in the right direction. Both of these will force you to simplify. And remember, simplicity is the ultimate sophistication.

2. Institute Simpler, More Effective Control Systems

In the past, systems have been developed to manage complexity. The result is complex systems. Today's success rate of managing with these complex systems and getting real, measurable, bottom-line results is poor. My opinion is the return on investment of the money and time put into manufacturing control systems is probably less than any other single business investment.

The complexity is not in the logic behind these systems—as a matter of fact, their logic is so simple it is often illogical. No, the complexity occurs because of the volume of data handled. Tens of thousands of parts, hundreds of work centers, thousands of vendors, and hundreds of distributors combine to require huge amounts of data for planning. Even larger volumes of input are needed to keep these plans current or warn of the need for replanning. Few people understand the details of these systems, simply because they are overwhelmed with the volume of data.

The move to simplify the environment, suggested above, will obviously reduce the amount of data needed. Removing problems will reduce inputs and be more effective than the systems approach designed only to help you cope with problems. But maybe the biggest gain with a simpler environment is that problems are made visible. The best on-line, real-time system is a pair of eyes and dedicated action. Now you get control without processing huge amounts of data.

This move to simpler systems will be a temporary respite, however. The automated factory of the future will need far more complex and integrated systems than we have today. Mathematical scheduling optimization, modeling, and simulation will all be necessary tools for the future. Operations research will finally be useful and start to earn its place in the arsenal of management weapons.

3. Spend More Time on Future Planning If You Want to Have a Future

It's a fact, your success or failure today depends heavily on the decisions made in the past. Yesterday's decisions on product development, machinery purchases, and market strategies determine your business performance today.

Even so, far too much senior management time is consumed fighting today's fires rather than planning for the future. Most are too busy keeping the ship afloat to sail it in the right direction. The time left for the future is totally inadequate.

Industry's rate of change is accelerating, however, demanding that you focus more on the future. Senior managers must look at least five years ahead to plot a corporate strategy that will succeed during the upcoming transition. And the strategic decisions must relate to the business in total, from vendor to factory to customer, and not be limited only to markets, products, and financing as they have in the past.

The risks of investing in the wrong technology, products, markets, or site location will become higher and the penalties greater. Future developments must be continually monitored, and strategy must be adjusted as events unfold. A failure to increase your dedication of effort to the future condemns your business to the past.

4. Give Your People the Right Training Now—Before It's Too Late

Manufacturing companies are operated by people. We hear lots of talk about robotics and the automated factory of the future, but the success or failure of these technological advances will still come down to people.

We have not developed our people to stay current with even today's level of technology. I'll leave you to consider what that means as we introduce tomorrow's.

From conventional planes to the Concorde: an aircraft analogy. In 1903 the Wright brothers were the first to sustain flight of a heavier-than-air plane. On their first flight they were airborne for 12 seconds. In 1927 Charles A. Lindbergh flew the Spirit of St. Louis from New York to Paris, a distance of about 3,000 miles, in 32 hours. In 1956 the Comet, the first jet airliner, entered regular service. In 1975, the Concorde started flying regularly from both London and Paris into Washington, D.C.

Each of these planes requires different skills to fly, even though the people flying them are or were all pilots. This difference in skills is especially evident in the case of the supersonic Concorde. Its engines are different from conventional commercial jets because of its afterburners; its flight characteristics are different because of the delta wing; its navigational system is different because of its speed (1,400 mph) and cruising altitude (10 miles); and its take-off and landing procedures are different because they use a "droop snoot" (the front hinged portion of the fuselage) to allow the pilots to see where they are going at slow speeds in congested traffic.

At times you fly the Concorde exactly opposite from a conventional wing plane. The Delta wing of the Concorde means that at slow speeds you must lift the nose of the plane very high for the wing to generate lift. If you tried this with a conventional plane, it would stall and crash. If you tried to fly the Concorde at slow speeds with its nose fairly low like conventional aircraft it would crash because of inadequate lift.

This analogy can be applied to manufacturing. Many of the tried-and-true concepts that worked well in earlier, less dynamic environments, will not work at all in today's business climate. In fact, they will be about as effective as flying the Concorde like a regular jet.

We expect and accept technology change in aircraft and train our flight crews, ground support personnel, and traffic controllers accordingly. The next advance in aircraft technology will likely be derived from the space shuttle. At that point, Concorde pilots, at the pinnacle of technology in today's commercial aircraft, will suddenly become obsolete. They will have to be retrained to learn how to fly this more technically advanced aircraft.

Manufacturing technology is changing faster than aircraft technology. In the field of operations planning and control, we have progressed faster than the aircraft industry. Extensive data handling and manipulation techniques became possible in the mid-1960s with the advent of powerful computer hardware. Software wasn't far behind, and the phenomenal proliferation of hardware and software since that time shows no signs of slowing down.

A handful of consultants and educators have preached the gospel of better control and pioneered many fine ideas in this field. The Just-in-Time methodology has shown us a degree of manufacturing perfection that few of us would have believed possible.

The problem is we have failed to train our pilots (senior managers), ground crew (operations people), and maintenance personnel (staff support groups). What training we have provided has been to enhance functional specialization. We make salespeople better at selling, engineers better at engineering, accountants better at finance, and manufacturing people better at production. Nowhere do we give these functional managers an overall view of the total process. In fact, our concentration on functional specialization reinforces the idea that a manufacturing concern can be effective when sets of experts address only their own unique areas. Nothing could be further from the truth. Suboptimization always results.

Industry needs an integrative process. Actions in one area have significant effects on others. These actions cannot be taken in isolation to improve a functional specialty without considering the repercussions on the business in total.

Logistics is an excellent way to get this company-wide focus. It shows clearly the interrelationships—why sales must relate to the factory, designers to process flow, manufacturing to market strategy, and so forth. Understanding these interrelationships, however, requires training in this total concept. Few companies are doing this effectively.

Industry must accept this training role. Most universities and colleges don't even address industrial subjects well, and those few that do are either fifteen years behind current technology and thinking or concentrate on functional specialization. What's the use of training people to be really effective at selling if they receive no training in what a manufacturer needs from salespeople to be effective? You can't learn this from being in the field, either. You need formal training in the total industrial process.

The people changes necessary to survive in the next ten years will be enormous. The time to start training your people is now—tomorrow will be too late.

5. Redefine Your Organizational Structure to Make Smart Decisions and Get Fast Results

Traditional industrial organizations are hierarchical in nature. They are modeled after the Roman army, not rigidly in groups of tens of course, but using many levels of management. Supremecy is at the top, with lower and lower levels of managers as you go down the organizational chart.

This is an effective organization for the military even today. Orders are unidirectional, from top to bottom. Little in the way of communication or advice is either expected or desired from the bottom up. But it is no longer an effective organization for industry. We need input from everyone in the organization to help beat the competition. Hierarchical organizations are notoriously poor for bottom-up communications because filtering, blocking, and altering of the information can be done by all intermediate levels. The result is frustration in the lower ranks and a lack of real facts at the top. More important than this, though, is that many-leveled hierarchical organizations communicate slowly in either direction. As the pace of change accelerates, this fact alone dooms the traditional hierarchical pattern.

Today's workers are ready and able to accept far more responsibility than we give them. Not only are they ready, they expect to control more of their day-to-day activities. This means a greater degree of freedom must be allowed or better yet, fostered. How do you do this with a hierarchical structure? Some companies are experimenting with or actually using matrix management where each manager reports to two business functions, for example, to a product line manager as well as to a unique resource manager. Although this works well in some companies, it clouds responsibilities in most. Matrix management is not very effective when key decisions have to be made and carried out fast.

Some Japanese companies have a better solution. They have attacked this problem by reducing the number of levels in the organization and expanding the scope of control. They generally have a much flatter pyramid in their organizational charts than comparable American or European companies. This obviously improves and speeds up communication in both directions and frees all managers and workers to accept a lot more responsibility.

The future will see more and more companies moving to small entrepreneurial teams. I call this a "round" organization. The leadership and makeup of these teams will be fluid and depend on the particular circumstances of the business at a point in time. The teams will be formed to manage projects, products, and divisions, and disbanded after the need has passed, only to be reformed again with a different group of players to tackle the next project.

These groups will be loosely tied to the corporate structure and given enormous autonomy. Success or failure will be completely in their hands, just as if they were running their own company.

These changes may be difficult to visualize today. The huge numbers of people in many industrial companies almost dictate hierarchy. But as the number of people in industry is reduced, as it will toward the end of this century, the round organization will become more common along with the systems technology to support it.

6. Pave the Way for Important Changes in Responsibility and Accountability

I devoted all of chapter 11 to operational performance measures. All of these are designed to push the organization toward an optimized total.

Traditional detailed financial measures so typical of most industrial concerns are going to be simplified. As direct labor costs drop more and more and the velocity of material flow increases, these measures will become less and less valid. Job costing will give way to process costing.

Responsibility and accountability assignment will become more difficult. Instead of focusing on prime responsibilities by individual or business function, we will switch to shared responsibilities reflecting the integrative nature of industry. In this environment, it will be critical to define the goals for the organization. These must first and foremost be operational in nature so that they will lead to good financial results.

Put on pressure for operational excellence. Develop your operating goals, get teams assigned to achieve them, and watch your financial results move in synchronization with improved operations.

BE SURE YOU REACH SIM BEFORE YOU TAKE ON CIM

The latest acronym in manufacturing companies is CIM, Computer Integrated Manufacturing. I prefer to call CIM Computer Integrated *Management* to ensure that everyone realizes it is a business process, not a factory process.

Many people are discussing how to implement this technology. They want to use it to beat the competition, foreign and domestic.

But CIM can also be spelled SIM. "SIM" stands for simplify. As mentioned in the simplified environment section, simple environments allow simple systems to be effective.

This concept is shown graphically in Figure 12-2. The complexity of CIM increases as you move toward a traditional, complex factory. It becomes much simpler as you move to a simpler factory, designated as a JIT operation. And because now CIM can be simpler, you will be able to introduce it quicker and earn benefits from it faster.

What Is CIM?

The best way to define CIM is with a diagram, as shown in Figure 12-3. Each circle contains a certain technology. The technologies are linked together by the lines.

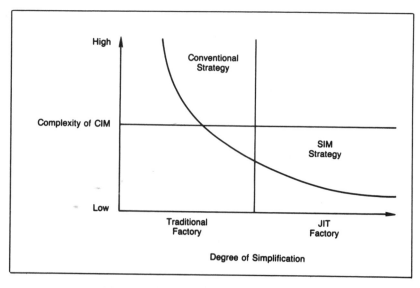

FIGURE 12-2. The Benefits of Simplification

We have been working toward this picture for at least 20 years. Many of the early developments did not consider integration, however, and resulted in the infamous "islands of automation."

At one point we called this process CAD/CAM (Computer aided design, computer aided manufacturing). One wag thought we called it this because it took two budgets to justify the expense. Because most of the progress was made in the CAD arena, another wag thought it should be called CAD/SCAM. Recently, however, the term CIM has begun to surpass CAD/CAM.

Nine Elements That Make up CIM

I will briefly define the contents of each circle and state the current level of technology.

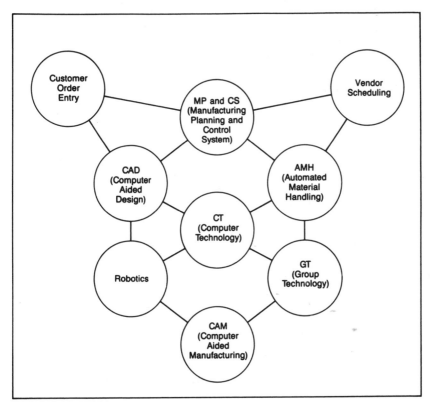

FIGURE 12-3. Computer Integrated Management

1. Electronic order entry speeds up customer orders. Many customer orders today are transmitted to the appropriate supplier by telephone or mail. Someone locally at the supplier company enters these into their control system. This is not true for many distribution systems however. Customers can enquire about the status of their suppliers' stocks by using a terminal and enter orders electronically.

This trend in distribution will accelerate and eventually will spill over into industry. Customers will then place orders directly on their suppliers, in fact booking against future planned production schedules. Maybe at some time, customers will even schedule their suppliers' plants.

2. CAD: "Aided" becomes "Automated." This is one of the most advanced technologies today. Wire drawings, three-dimensional solids modeling, and finite element analysis are all already well developed and currently in use.

The trend is toward removing the word "aided" and replacing it with "automated." This requires defining the design criteria and letting the computer detail the product based on customer specifications. This is already possible for some products. As an example, many kinds of power transformers are automatically designed today based on the customer specifications. The computer uses these specifications to create the exact details of how to build the product. The same is true for some highway trailers. Other products, such as printed circuit boards and custom semiconductors, are fast approaching this condition.

3. Robotics technology advances beyond pick and place. With today's level of technology, robots are largely pick and place. They move in a prescribed way with little tolerance for part location or position.

Vision systems, still largely under development, are needed to provide tolerance of part location and orientation. Vision will also allow robots to travel independently in three dimensions.

4. MRP II gives way to schedule simulation and optimization. MRP II is today's state-of-the-art scheduling and control system. It will change dramatically as we automate plants. MRP II has many failings. For example, it is unidirectional in information flow, it is a backward scheduling process, it assumes infinite capacity, parameters flexible in the business are fixed in the system, it cannot automatically deal with alternate materials or process routings, and it doesn't judge the quality of its decisions.

These failings will all be overcome through schedule simulation and optimization. Artificial intelligence will select the best course of action from various alternatives. The selected option will be constantly updated and reoptimized, not periodically updated and replanned using huge people inputs and review, as is the case today.

5. *Advanced computing technology gives greater depth to calculations and communications.* Just becoming effective are relational data bases and fourth generation languages. The Manufacturing Automation Protocol (MAP), spearheaded by General Motors, will allow different makes of software and hardware to communicate with each other. Artificial intelligence and expert systems will add judgment and experience to simple numerical calculations.

6. *Computer aided manufacturing expands to include computer aided process planning.* This segment of CIM encompasses all the instructions needed to operate the factory machinery. Right now CAM consists largely of numerical controls (N.C.) tape preparation and computer numerical control (CNC) programs along with N.C. and CNC machine operation. This circle will be expanded in the future to include computer aided process planning. The specific routing of parts and the operational instructions for automated machines will be generated at first with human help. Complete automation will come later.

These activities will become an integral part of the scheduling system. Process routings and schedules for a part will be selected based on capacities and existing workloads of machines. The operational instructions for machines will be tightly linked to the operational instructions for robots and automated guided vehicles.

7. *Automated material handling becomes more sophisticated.* Driverless tractors and automated storage and retrieval systems are both well-developed technologies. Unfortunately, at today's level of expertise, we can move and store all the wrong things very efficiently.

The role of automated storage and retrieval will diminish as customers and suppliers integrate their schedules. Just-in-Time will really be possible at that time. More freedom of pathways will be given to automated guided vehicles rather than the embedded wire or optical tracing systems currently in use.

8. *Group technology undergoes radical changes.* This circle encompasses two processes. First, coding and classification systems that identify design and process similarities in parts. Second, the organization of groups of dissimilar machines into production cells.

This first process, classification and coding, will become an internal computer process, aided by artificial intelligence. The second, cell groupings, will largely disappear. They do not form the most flexible layout. Cells are created largely to accommodate humans so that material handling is reduced and scheduling is much simpler.

With automated guided vehicles and an optimized scheduling system, the two primary reasons for cells disappear. Machinery will then be located in certain parts of the factory for other reasons, such as maintenance ease or environmental restrictions.

9. *Customers take an active role in vendor scheduling.* As we've discussed, most customer orders are currently placed with vendors by using the mail or telephone. The vendor's schedulers then decide when to promise delivery, and they do their own internal scheduling.

In the future, we will connect the customer's scheduling systems directly with the vendor's. At first this will be done by booking orders directly against the vendor's upcoming production schedules or open capacity. Later, customers will actually schedule their vendors' operations within established guidelines.

Using New Benefits to Justify CIM

Today, we use classical cost justification methods, including return on investment, discounted cash flow, or internal rate of return to decide whether to invest in a new system or piece of machinery. If the system or piece of equipment exceeds a given threshold rate of return, we usually get approval to go ahead.

CIM will not be justified the same way, for a variety of reasons. First, it is an enormous project, probably without end. Today's technology will be replaced with tomorrow's, tomorrow's with technology developed the day after, and so on in a never ending cycle of renewal.

Second, because of its size, CIM will have to be installed in pieces. Many of the benefits, though, won't come from installing a given piece of technology but from the synergy of integrating several pieces together.

Third, most of the benefits will not come from tangible, post-auditable results. These are what I call "hard" benefits. Instead, many of the benefits will be from difficult to qualify benefits, in essence, "soft" benefits.

Seven "Soft" Benefits from CIM

1. *Greater flexibility.* The whole thrust of the new technology is flexible automation. This means the ability to make a variety of prod-

ucts easily on a piece of equipment simply by changing software. Compare this to earlier forms of automation where products are hard-tooled. Changes are difficult if not impossible with this kind of equipment.

2. *Shorter, more reliable customer lead times.* The reason for long lead times to customers is due in large part to the nonintegrative nature of our businesses. Salespeople, engineers, schedulers, factory supervisors, and vendors all need time to process and relay information to one another.

Scheduling difficulties, poor quality, machine breakdowns, and the like make keeping promises to customers almost impossible.

Integration of information will speed up the time it takes to process a customer's order. Highly reliable and optimized schedules will ensure that promises are kept.

3. *Introduce new products faster to stay competitive.* Many companies succeed or fail based on their ability to introduce new technology or new products quickly. This "quick-to-market" capability could be for offensive reasons—to beat the competition with new products—or for defensive reasons, to catch up with the competition.

Like shorter, more reliable lead times, this benefit of CIM will come about from electronic technology, integration, better scheduling, and highly reliable execution.

4. *Few "fix it" engineering changes.* Today's computer aided design and computer aided engineering systems subject new products to many tests while the product is still on the drawing board. Tolerance buildup of parts, functional operation, strength, and deflection estimates can all be simulated electronically. This ability to test a product thoroughly in the design stage can eliminate many corrective actions needed later.

5. *Better product quality.* Product quality is affected by many factors. One is the design itself, which can be improved as mentioned above. Another is the process, which will become more reliable as automation reduces the human element. Still another is the purchase of raw materials or parts, which will be improved because specifications will be better communicated from customer to supplier.

6. *Empty floor space permits new or expanded product lines.* Automated factories will be much more compact than conventional plants for several reasons. First, machines will perform multiple operations. Second, scheduling will reduce the need to store inventories.

At first glance—especially if you have a financial background—you may question why empty floor space is a benefit. And if it stays empty, it certainly isn't.

But if empty floor space allows you to bring some new products into the factory or expands production of existing products without adding bricks and mortar, then it is very much a benefit. And a corollary is that new plants will be much smaller than traditional plants, allowing a much higher asset turnover.

7. *More flexible, better trained workers.* Operating tomorrow's factories will require far more skills and knowledge in all our people. Think of the Concorde analogy we used earlier. A pilot of a single engine plane flies mainly by looking out the window. His flight instrumentation is minimal. His is largely a manual skill. The Concorde pilot doesn't bother to look out the window except when taking off and landing. What's the point? If he saw something, at 1400 mph it would be too late.

Instead, he relies on computers, instruments and instructions from air traffic controllers. His is a knowledge-based skill. He uses his manual skills only in cases of emergency.

The same will be true of people in industry. They will make the transition from operating simple machines manually or performing simple clerical tasks to monitoring groups of complex machines or using simulation to decide on the best way of solving complex problems. Making your people more knowledgeable could be the most important benefit of all.

Don't Overlook the Hard Benefits

CIM as described in Figure 12-3 is not technically feasible today except in some limited types of manufacture. But the missing pieces of technology are all under active development. CIM *will* be technically feasible by the year 2000.

Along with the soft benefits, significant hard benefits will accrue to those who implement CIM. By operating discrete manufacturing plants around the clock as some process industries do today, you will be able to realize enormous advantages over current operating times, shown in Figure 12-4. These numbers are a little old but they haven't changed much since 1980. The 8 percent productive cutting number can be as high as 96 percent in a continuous flow process. (I doubt a discrete manufacturing plant could achieve 96 percent even fully automated, but at least it wouldn't have to shut down on second or

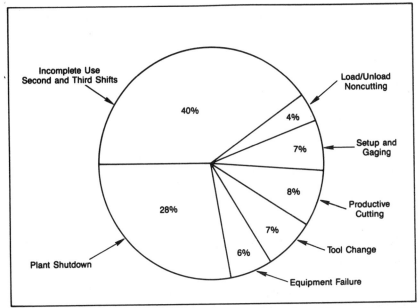

Source: *American Machinist Special Report, October 1980*

FIGURE 12-4. Breakdown of Production Time

third shifts, weekends and holidays. That should multiply the 8 percent by 3 at least.) Capital asset turnover will be much higher, and don't forget the potential inventory savings from integration.

Even so the biggest benefits will still be the soft ones. Flexibility, shorter, more reliable lead times, higher quality, and better people will make you (or your competition) exceedingly competitive.

A Four-Step Action Program to Help Your Company Get to CIM

The new technologies embodied in CIM show a lot of promise. No one knows for sure exactly how they will develop or if the fully automated factory is economically viable. Even so, some pieces of the technology have already proved to be beneficial beyond question. Others still under development will also have excellent payback. You can start taking advantage of the new technology with this four-step program.

Step 1: *apply today's state-of-the-art technology*. Many elements of CIM are technically possible today. Depending on your particular circumstance, these technologies, even at today's level of develop-

•	5– 20%	Reduction in Personnel Costs
•	15– 30%	Reduction in Engineering Design Costs
•	30– 60%	Reduction in Overall Lead Time
•	30– 60%	Reduction in Work-in-Process
•	40– 70%	Gain in Overall Production
•	200– 300%	Gain in Capital Equipment Operating Time
•	200– 500%	Product Quality Gain
•	300–3500%	Gain in Engineering Productivity

National Research Council

FIGURE 12-5. Benefits of CIM

ment, can give excellent payback. Figure 12.5 shows the results of a study conducted by the National Research Council on the measured benefits from CIM. As you can see, some of the results are outstanding.

This means you should make every effort to install the current technologies that will benefit your business. Make sure you give adequate credit for the soft benefits, and make sure you implement SIM before CIM. Simplify your environment, using logistics as your main focus. Then your company will find it easier to inaugurate CIM concepts simply and effectively.

Step 2: stay abreast of developing technologies. Create a formal program to keep you up-to-date with CIM developments. This program should be headed by a senior manager. You'll need someone with a technological background because tomorrow's factories will be technology based. You'll need a senior manager to give this project adequate recognition and to ensure an entrepreneurial slant.

Set up a structured data gathering team. Because CIM is such a huge project, team representatives should come from every major business discipline. Members will focus on developments in their particular areas.

Collect data systematically. This can be done internally, by subscribing to automation magazines, or visiting trade shows and conferences. It can also be done through an outside consulting service specializing in collecting and screening data about automation subjects.

Make sure the data gathering is worldwide in scope. Many countries have joined the race toward a fully automated, discrete products factory. If you focus only on domestic authorities, you will limit your scope too much.

Prepare a formal schedule for reviewing the status of technology, perhaps quarterly. Make sure your findings are transmitted throughout the organization, especially to senior managers. This will keep them abreast of evolving technology and how it can be used to benefit the business.

Step 3: implement promising ideas. Some of the new techniques you identify in your review process will show promise. But it will be impossible to guarantee results. The uncertainties associated with untested, synergistic technology will prevent it. Hence, you have to accept a certain amount of risk. Implementing CIM will be much like an R & D program, except that you are experimenting with advanced ways of operating your business.

Make sure you select technologies with integration as a long-term objective. Nonintegral solutions are acceptable only if they have a fast payback. That way, you can afford to discard them if integrating proves too difficult later on.

Assuming normal payback periods, however, consider integration as you select areas of CIM to address. Ensure that the specifics of your selected area are compatible with existing pieces of the CIM diagram and will remain compatible in the future.

Step 4: develop tomorrow's skills today. The major problem in getting useful results from the coming technologies will be the lack of sufficiently qualified people to manage them. This deficiency will be a problem at all levels of the organization, from the board room to the lunch room.

Industry must accept the role of training its existing work force to become qualified to run the factory of the future. Universities and colleges are not geared up to retrain large numbers of existing employees; nor will government training programs be sufficient. Both tend to train people in today's or yesterday's skills. We need tomorrow's skills today.

A handful of companies are already doing a good job of training their people. A division of 3M spent 4 to 5 percent of the working hours of its people in formal classroom training in 1984 over and above on-the-job training. By 1986, its employees spent an average of 7 percent of their working hours in formal classroom training.

However, even 7 percent may not be enough to keep up with new developments. Thomas Gunn of Arthur Young predicts that, by 1995, the majority of your work force must spend 20 percent of their working hours in formal education or they will not be able to keep up with evolving technologies.

I don't know if 20 percent by 1995 is correct. I do know that today's typical attitude toward education is deficient. Employee education is poorly budgeted, probably averaging about 1 percent of working hours, and it's chopped off as soon as business slows down. It is considered a "deferrable expense."

This attitude has to change. Many senior managers give lip service to the idea that "People are our most important asset." But as soon as money gets tight, they stop investing in their most important asset. Deterioration immediately sets in.

Education and training of the work force has to be planned, budgeted, and implemented through thick or thin. Only then will you get people who are qualified to lead you successfully into the twenty-first century.

YOU'VE GOT TO GET MOVING TO GET THERE FIRST

The factory of the future is fast approaching. We may get there in the next ten years. The company that successfully masters this new technology will have an unparalleled advantage over the competition. Speed of response, quality of product, design excellence and costs will all be significantly improved.

Getting to this advanced condition will take creativity, an acceptance of risk, and a drive to get there first. But above all, it will take people with a vision of the utopian factory, qualified enough to get you there.

Bibliography

Abegglen, James and George Stalk, *Kaisha*, Basic Books, 1985.

Goldratt, Eliyahn and Jeff Cox, *The Goal*, North River Press, 1984.

Gue, Frank, *Increased Profits Through Better Control of Work in Process*, Reston Publishing Company, 1980.

Hall, Robert, *Attaining Manufacturing Excellence*, Dow Jones Irwin, 1987.

Johnson, Thomas and Robert Kaplan, *Relevance Lost*, Harvard Business School Press, 1987.

Porter, Michael, *Competitive Strategy*, The Ree Press, 1980.

Shingo, Shigeo, *A Revolution in Manufacturing: The SMED System*, Productivity Press, 1985.

Shingo, Shigeo, *Study of Toyota Production System*, Japan Management Association, 1981.

Skinner, Wickham, *Manufacturing in the Corporate Strategy*, John Wiley and Sons, 1978.